Food and Livelihood Security

SIKSHA 'O' ANUSANDHAN
(Deemed to be University), Bhubaneshwar-751030 Odisha

NIPA® GENX ELECTRONIC RESOURCES & SOLUTIONS P. LTD.
New Delhi-110 034

Food and Livelihood Security
Challenges and Opportunities

Edited by

N.K. Dhal
Adjunct Professor (Plant Pathology)
Institute of Agricultural Sciences, SOA (Deemed University)
Bhubneshwar, Odisha
and
Former Dean, College of Agriculture
Odisha University of Agriculture and Technology, Bhubneshwar
& SOA (Deemed University), Bhubneshwar, Odisha

P. Mahapatra
Former Dean Post Graduate Studies
Institute of Agricultural Sciences, SOA (Deemed University)
Bharatpur, Bhubaneswar, Odisha 751003

B.K. Sahoo
Advisor and Former Dean
Institute of Agricultural Sciences
SOA (Deemed University), Bhubaneswar

S.K. Mohanty
Head
Crop Physiology
Institute of Agricultural Sciences
SOA (Deem University), Bhubaneswar

SIKSHA 'O' ANUSANDHAN
(Deemed to be University), Bhubaneshwar-751030 Odisha

NIPA® GENX ELECTRONIC RESOURCES & SOLUTIONS P. LTD.
New Delhi-110 034

SIKSHA 'O' ANUSANDHAN
(Deemed to be University), Bhubaneshwar-751030, Odisha

NIPA® GENX ELECTRONIC RESOURCES & SOLUTIONS P. LTD.
101,103, Vikas Surya Plaza, CU Block
L.S.C.Market, Pitam Pura, New Delhi-110 034
Ph : +91 11 27341616, 27341717, 27341718
E-mail: newindiapublishingagency@gmail.com
www: www.nipabooks.com

For customer assistance, please contact
Phone: + 91-11-27 34 17 17
Fax: + 91-11- 27 34 16 16
E-Mail: feedbacks@nipabooks.com

ISBN: 978-81-19002-42-9

Composed and Designed by NIPA®.

SIKSHA 'O' ANUSANDHAN

(A Deemed to be University Declared u/s 3 of UGC Act, 1956)
Re-Accredited by NAAC with A' Grade

Prof. (Dr.) D.P. Ray
FNAAS, FNABS. FHSI, FISVS, FCHI, FAIPUB, FRFRI, FOES
Former Vice-Chancellor, OUAT
Chancellor

Foreword

It gives me immense pleasure that Siksha 'O' Anusandhan is publishing the book entitled "Food and Livelihood Security - Challenges and Opportunities" in collaboration with NIPA, New Delhi. The articles were based on technological innovations, policy planning and interventions in order to achieve food, nutritional and livelihood security keeping in view the challenges ad opportunities in the field of agriculture and allied sectors.

Agriculture production in recent times is declining in spite of technological advancements in agriculture paving the way to think and act differently in the context of climate aberrations and consequent effect on income of farmers. Currently the world is facing great challenges in agricultural sectors. Climate change and natural calamities adversely affecting agricultural production; malnutrition among the children within the age group of 5 years is alarmingly increasing affecting brain development and the recent Ukraine - Russia war has completely disrupted the supply chain of food grains in the world. This book has included futuristic concepts made by eminent agricultural scientists of India in different thematic areas Viz. technological innovations, capacity building and technology transfer, entrepreneurship development, organic farming, agricultural financing, challenging issues in order to achieve nutritional and livelihood security. The authors have contributed articles based on research, policy planning needs and critical technological interventions needed to achieve the goals.

1 congratulate the editors for rightly selecting the articles for the publication in the book and believe, the articles will definitely help the researchers, students, planners and policy makers to think on the issues related to sustaining food and livelihood security of our farmers.

20/8/2022

Prof. (Dr) D.P Ray
Chancellor

Preface

Food, an essential commodity of our daily diet is the united effort of millions of farmers, agricultural scientists, traders and administrators from all over the world. In India, Agriculture is the biggest industry in term of providing employment to 55% of the population. It is the most efficient user of the solar energy that produces food and supplies oxygen for life; besides mitigating the climate change impacts. During COVID-19 pandemic in 2020-21; the growth of agriculture sector accounted for 20% to GDP in comparison to the industrial sectors which showed negative trend of growth, particularly small to medium scale industries were severely affected. Reverse migration to rural hometowns seems to be rising among Indian youth's, plagued with post Covid-19 employment uncertainty and alienation. The promise of a better lifestyle coupled with technology transformation in agriculture has opened new possibilities.

Agriculture presently is standing on a cross road. In one hand we have to produce more to meet the demand of constantly growing population and on other hand many adverse conditions like increased events of natural disasters, decline in factor productivity, shrinking cultivable land due to increasing urbanization, fragmented land holdings, etc. have posed newer challenges towards the agricultural growth in India. According to the survey made by NABARD, the agriculture households showed a greater dependence on multiple source of income, with 50% reported to have two sources of income, while closer to 40% depend on more than two sources for income. It indicates that income only from agriculture is not enough to meet the household needs. Therefore, for production enhancement in agriculture and increase in farm income, we have to make situation based strategic planning by combining the advanced technologies with farm mechanization, smart agriculture and promoting agripreneurships.

Siksha 'O' Anusandhan DU has rightly organized the "National E-Conference" at its Institute of Agricultural Sciences to explore the avenues in agricultural sector with regards to technological innovations, employment generation through capacity building and agripreneurship, improving agricultural marketing systems, emerging challenges due to climate changes and over all on nutritional security for sustainable livelihood. For an agrarian based economy like India, agricultural sector plays an important role as its growth and livelihood are inter dependent.

"Food and Livelihood Security- Challenges and Opportunities" develops an understanding of the diversity and the complexity of agriculture sector. The authors of individual chapters have endeavored to present the state of the art in their areas of expertise, pointed out areas where new work has yielded exciting results and/or controversies. The eminent scientists and the policy makers have presented their personal opinions and put forth the direction wherein the future research should proceed.

I wish to express special thanks to all the members of the Editorial Board for their energetic devotion and untiring effort in preparation of this manuscript. On behalf of the Editorial Board, it is my privilege to acknowledge the unconditional support and help extended by Prof. Manojranjan Nayak, the founder President, SOA, a renowned educationist and visionary and Prof. D. P. Ray, Chancellor, SOA. I thank each author for the enthusiasm, consistency and patience to accomplish the present assignment.

Editors

Contents

List of Contributors

A.K. Behera, Former MD, OSIC, OAIC & APICOL., Bhbaneswar

A. Mishra, PG College of Agriculture, Dr. Rajendra Prasad Central Agricultural University, Pusa, Samastipur, Bihar

A. Tiwari, Baba Raghav Das Post Graduate College, Deoria, RD Marg, Uma Nagar, Mehra, Uttar Pradesh

A.V. Kothiya, Department of Farm Machinery and Power Engineering, College of Agricultural Engineering & Technology, Anand Agricultural University, Godhra, Gujarat

B.B. Priya, Department of Soils and Environment Agriculture College and Research Institute (TNAU),Madurai, Tamil Nadu

B.K. Dadarwal, Institute of Agricultural Sciences, Banaras Hindu University, Uttar Pradesh

B. Parasar, Professor, Department of Extension Education, Siksha 'O' Anusandhan Deemed University, Bhubaneswar, Odisha

D. Dash, Department of Soil and Water Conservation Engineering, CAET, Odisha University of Agriculture and Technology, Bhubaneswar, Odisha, India

D.K. Dwivedi, Faculty of Agriculture, Sri Sri University, Cuttack, Odisha

G.C. Acharya, Central Horticultural Experiment Station, ICAR-IIHR, Aiginia Bhubaneswar-751019, Odisha

G. Sahoo, Krishi Vigyan Kendra, OUAT, Angul, Odisha

J.M.L. Gulati, Professor, Department of Agronomy, Institute of Agricultural Sciences (SOA DU), and Former Dean, Post Graduate Faculty- cum- DRI, OUAT, Bhubaneswar, Odisha

K.B. Mohapatra, Professor, Department of Plant Pathology, GIET University, Gunupur, Odisha

K.S. Giri, Assistant Professor Deparment of Agronomy, Institute of Agricultural Sciences, Siksha'O'Anusandhan (Deemed to be University), Bhubaneswar, Odisha

K. Sar, Assistant Professor Department of Agronomy, Faculty of Agricultural Sciences, Siksha 'O' Anusandhan, Bhubaneswar, Odisha

L. Das, Principal Scientist & Nodal, AICRP (Women in Agriculture), ICAR-CIWA, 7QPP+584, Kalinganagar, Bhubaneswar, Odisha

L.N. Gediya, Department of Genetics and Plant Breeding, B. A. College of Agriculture, AAU, Anand, Gujarat

I.P. Pattnaik, Department of Extension, Institute of Agricultural Sciences, SOA (DU), Bhubaneswar, Odisha

M. Mishra, Department of Agricultural Economics, College of Agriculture, Odisha University of Agriculture and Technology, Bhubaneswar, Odisha

M. Sahoo, Department Extension Education, Institute of Agricultural Sciences, Sikhsha 'O' Anusandhan, Bhubaneswar

N. Jena, Associate Professor Department of Plant Physiology, Institute of Agricultural Sciences, SOA DU, Bhubaneswar, Odisha, India

N.K. Dhal, Adjunct Professor (Plant Pathology), Institute of Agricultural Sciences, SOA (DU) and former Dean, OUAT & SOA, Bhubaneswar, Odisha

N.N. Mishra, Agriculture & Inclusive Banking, DCB Bank, Bhubanewar, Odisha

N. Pradhan, Department of Agronomy, Odisha University of Agriculture and Technology, Bhubaneswar, Odisha

N. Yadav, Department of Horticulture, Institute of Agricultural Sciences, BHU, Uttar Pradesh

P. Mahapatra, Former Dean,Post Graduate Studies, Institute of Agricultural Sciences, SOA (DU), Bharatpur, Bhubaneswar, Odisha 751003

P. Sahu, Department of Vegetable Science, OUAT, Bhubaneswar, Odisha

S.C. Senapati, Institute of Agricultural Sciences, Siksha 'O' Anusandhan, Bhubaneswar, Odisha

S.J. Jena, Agricultural Extension and Communication, Institute of Agricultural Sciences, Siksha 'O' Anusandhan University, Bhubaneswar, Odisha

S.K. Behera, Professor & Head, Dept. of Agricultural Engineering, SOA DU, Odisha

S.K. Dwibedi, Odisha University of Agriculture and Technology, Bhubaneswar, Odisha

S.K. Mahapatra, Faculty of Agriculture, Sri Sri University, Cuttack, Odihsa

S.K. Panda, Dean, School of Agriculture, GIETU, Gunupur, Odisha

S.L. Babu, Department of Agricultural Extension, College of Horticulture, Chinalataripi. Dr.Y.S.R. Horticulture University, Hyderabad

S.S. Nanda, Professor, Department of Agronomy, Institute of Agricultural Sciences, SOA (DU) Bhubaneswar, Odisha

S. Shasani, Department of Agricultural Extension and Communication, IAS, SOA (DU) Bhubaneswar, Odisha

U. Das, Gandhi Institute of Engineering and Technology University, Gunupur, Odisha

1

Smart Agriculture A Game Changer for Farmer's Welfare

S.K. Panda and S. Panda

Introduction

Smart farming focuses on retrieving data and using it in developing farming systems in a smart way that will yield better results. Such real time data can be readily accessed by the farmers through smart phones or tablets which will help them to take wise decisions at the right time. It helps in managing farms using technologies like IoT, robotics, drones and AI to increase the quantity and quality of products while optimizing the human labor required by production. Digital farming goes a step further and creates value from data. Integrating the concept of precision and smart farming, digital farming can be explained as "consistent application of the methods of precision farming and smart farming, internal and external networking of the farm and use of web-based data platforms together with Big Data analysis" (A paper on Digital Agriculture, 2018). Though such technologies are highly attractive and worth investing, they are also accompanied by uncertainty and trade – offs. Education, training, dissemination of information, finance and marketing are some of the major factors that would drive the farmers towards adoption of sustainable farming technologies (Grand View Research, 2019). Precision farming uses techniques to precise the amount of inputs into the farming system considering their impact on environment. Following the two waves of revolution in agriculture namely mechanization and green revolution, precision farming is now considered the third wave and is still being enhanced using digitalization process (Zhang, 2019). Being the largest user of land, agriculture as a production unit has been facing great challenges. The new and progressive concepts in agriculture have to be highly location specific which will satisfy the biodiversity of that area and need of the situation. Digitalization here, assumes a prime role which will help in maintaining value creation in agriculture by modifying the farming and business culture.

Pre-requirements for Digitizing Agriculture

1. Improving facilities for telecommunication: This has to be done especially in rural areas so that the industry/market can be directly linked with the farm.
2. Ensuring data security: The data collected by the farmer through regular monitoring has to be secured and regulations must be imposed to prevent illegal usage of such data, however, maintaining transparency in the entire process.
3. Preventing uncontrolled transfer of data to cloud systems: The base data belongs to farmers and they should be benefitted for the same. Avoiding data monopoly will restrict random public and third party usage, ensuring sound functioning of the digitizing system.
4. Provision of central data portal: Though usage of third parties are to be restricted, yet, the data (weather data, soil data, water data, land register data, route networks, etc.) should be kept in open access format (viewing purpose), since all the data collected and archived are financed by taxes.
5. Bringing agriculture and consumers closer to each other: Transparency and production method traceability are key points in digitization. This creates confidence in producers and establishes trust in consumers.
6. Qualification before digitization: Since digitizing agriculture is a sensitive process, the persons involves in the process, including farmers, must be well trained and skilled for the process.

The concept of digitization to minimise labour and increase production efficiency is not new. Some well-established digitally assisted agricultural equipment are already in use as in sensors, self- steering machines, variable rate application machines for fertilizer and pesticides, combine harvesters, milking machines, automated feeding and manure removal robots. The successful adoption of such technologies elevates the importance of digitization in agriculture improving production, efficiency and sustainability, that too, through eco-friendly tools. This will ensure low yield losses, high quality produce and much reduced on/ off farm impact on the environment.

Internet of Things (IoT)

IoT in agriculture refers to use of sensors and other devices to transform farm activities and status into a set of data. IoT technologies hold the potential to generate a lot of valuable information in agriculture and helps in the process of digitalization of agriculture helping in achieving higher crop yield with lesser cost. By 2050, the agriculture sector needs to produce 60% more food globally to ensure food security and IoT, if utilised aggressively now, can increase agricultural productivity by 70 % by 2050 (Sarni et al.). The benefits that the farmers are getting by adapting IoT are twofold. It has helped farmers to decrease their costs and increase yields at the same time by improving farmer's decision making with accurate data.

Applications of IoT in Agriculture

1. Enables us to know real-time weather conditions through sensors: It is highly essential for the farmers to take time specific decisions by the farmers for their farms.
2. Enables the farmers to adopt smart farming: Soil and water management, livestock monitoring, vehicle tracking, field observation and inventory monitoring.
3. Automated/ Smart greenhouses: The in-house climatic conditions detected by sensors and are adjusted as per instruction fed to the equipments, significantly reducing human intervention, thus making the whole process cost effective.
4. Better interpretation of data: Since data collected by IoT driven sensors are accurate, the analysis and interpretation of the values too, will be precise. Trend analysis of such data can help in forecasting and planning the next action.
5. Drones: Over viewing the field, without actually getting into it. Drones will make monitoring more efficient. They can also be used for spraying purpose.

The applications of IoT-based smart farming not only target conventional, large farming operations, but could also act as new levers to uplift other growing or common trends in agricultural like organic farming, family farming (complex or small spaces, particular cattle and/or cultures, preservation of particular or high-quality varieties, etc.), and enhance highly transparent farming. In terms of environmental issues, IoT-based smart farming can provide great benefits including more efficient water usage, or optimization of inputs and treatments.

Artificial Intelligence and Robotics

Artificial intelligence (AI) is a branch of computer science research whose goal is to build computer programs in order to understand intelligence and exhibit intelligent behaviour. AI based technologies in agriculture manage challenges related to crop yield, irrigation, soil content sensing, crop monitoring, weeding and crop establishment (Kim et al., 2008). AI, like IoT can be a solution to the crisis being faced due to soaring of global population. Such technologies are used in precision farming, enabling the farmers to produce more of better quality output and with less input, which in turn gets well accepted by consumers and gets established in market.

Applications of Artificial Intelligence

1. Unmanned Aerial Vehicles (UAVs): Because of their versatility as well as amazing imaging technology which covers from delivery to photography, the ability to be piloted with a remote controller and the devices being dexterous in air enables us to do a lot with these devices. They are becoming increasingly popular to reach great heights and distances and carrying out several applications in agriculture.

2. Reduction in workforce: AI helps make farming easier by mixing technology with biological science. Minimizing the workload of farmers, AI makes the process of data collection much accurate than manual monitoring, reducing the labour to a great extent. AI in agriculture can be applied which would automate several processes, reduce risks and provide farmers with a comparatively easy and efficient farming.

3. Chatbots for farmers: AI accompanied by machine learning can yield revolutionary changes for the farmers. Chatbots have been designed to interact with farmers in a more personalised way, assisting them in receiving answers to their queries and providing them recommendations for the same.

4. Expert system (ES) and Decision Support System (DSS): The combination of IoT and AI in the form of ES and DSS is a blessing to the farmers. The Agriculture Expert Systems are designed to emulate the logic and reasoning processes that an expert would use to solve a problem. They are based on the integration of knowledge and experience of specialists from different fields and have the capability to answer relevant questions and explain its reasoning process. Such systems can be developed and applied by scientists and farmers in various fields of agriculture like plant breeding, agronomy, plant pathology, entomology, agricultural extension, marketing agricultural commodities, financial decision making, economic and entrepreneurship development in agriculture. DSS are software-based systems that gather and analyse data from a variety of sources. Their purpose is to smoothen the decision-making process for management, operations, planning, or optimal solution path recommendation. In the agricultural sector, it helps farmers to solve following complex issues related to crop production.

1. Pest and disease management
2. Nutrition management
3. Irrigation scheduling
4. Crop selection
5. Land usage
6. Breed selection
7. Market selection
8. Crop yield prediction
9. Climate change adaptation
10. Food Waste Control – selection of supply chain

Machine learning is the study of computer algorithms that allow computer programs to automatically improve through experience. It is emerging together with big data technologies and high-performance computing to create new opportunities in precision agriculture. Some of the practical applications of machine learning are:

1. Species Recognition/management
2. Soil management
3. Water Management
4. Yield Prediction
5. Crop Quality
6. Disease Detection
7. Weed Detection
8. Livestock management

Robotics has played a substantial role in the agricultural production and management. Since the conventional farming techniques lack in efficiency, researchers have started emphasizing on applying AI in agriculture through robots (Dursun and Ozden, 2011), which would replace human labour and produce effective benefits on small as well as large scale productions (Manivannan and Priyadharshini,2016). The robots are performing various agricultural operations autonomously such as weeding, irrigation, guarding the farms for delivering effective reports, ensuring that the adverse environmental conditions do not affect the production, increase precision, and manage individual plants in various unfamiliar ways.

Application of Robotics

1. Irrigation: The technology of smart irrigation is developed to increase the production without the involvement of large number of man power by detecting the level of water, temperature of the soil, nutrient content and weather forecasting. The actuation is performed according to the microcontroller by turning ON/OFF the irrigator pump. The water will be supplied by the resource according to the requirement and it will also update and store the sensor values.
2. Weeding and spraying: Before developing a weed control automated system we need to differentiate between the crop seedlings and the weeds. The vision based technologies are used to guide the robots along the row structure to remove weeds and to differentiate the single crop among the weed plants. Similarly, automated machines can be used for efficient spraying over crops thus minimising the labour requirement.
3. Drones: Technological advancements has almost revolutionized the agricultural operations and the introduction of agricultural drones is the trending disruption. The ground and aerial drones are used for assessment of crop health, crop monitoring, planting, crop spraying, and field analysis. With proper strategy and planning based on real-time data, drone technology has given a high rise and makeover to the agriculture industry. Drones with thermal or multispectral sensors identify the areas that require changes in irrigation. Once the crops start growing, sensors indicate their health and calculate their vegetation index. Eventually smart drones have reduced the environmental impact. The results have been such that there has been a massive reduction and much lower chemical reaching the groundwater.

Geospatial Technologies

The data collected by the farmers and officials need to be fed to the databases and analysed effectively for proper applications of geospatial technologies in agriculture. For this purpose, many advanced techniques have been invented and utilised which use the principle of remote sensing (RS), that divides large farms into management zones and the requirements of each zone vary, which are studied through Global Positioning System (GPS) and Geographic Information System (GIS).Most farms are generally considered to have records of soil survey, history of crops, cropping practices of the region and these data can be supplemented with aerial and satellite photographs during different period of the year or seasons. This will enable the farmers to identify issues of different management zones and accordingly decide the inputs restricted to that area. Since such data can give an idea to the farmer about the productivity status of the zones, the growth and yield patterns of respective zones from the farm can also be identified.

The observations made through the geospatial technologies are geo-referenced into a GIS database where it will be further analysed and interpreted. Image-based data captured through RS accompanied by AI, can be beneficial to agriculture in many ways:

1. Crop identification
2. Crop acreage estimation
3. Crop condition assessment and stress detection
4. Identification of planting and harvesting dates
5. Crop yield modeling and estimation
6. Identification of insect pest and disease infestation
7. Soil moisture estimation
8. Irrigation monitoring and management
9. Soil mapping
10. Monitoring of droughts
11. Land cover and land degradation mapping
12. Identification of problematic soils

Apart from direct involvement of RS in agriculture as above, it collects some point data, which gets stored and analysed for them to be useful for the farmers. These point data get converted to spatial data, reflecting the entire farm. Following this, the GIS tools and software can be used to develop digital maps wherein the ground data (spatial information) collected is converted to digital format. Thereafter, imaginary grids are developed, where points with different values are represented with different colours. Assigning colours to digital maps makes it easier for the user to identify the values, compare and interpret them. Thus, with such technology in use, suitable farming techniques can be used by farmers as per zones and take up site specific decisions.

Applications of GPS

1. High resolution surveying
2. Grid sampling

3. Yield mapping
4. On the go sensors
5. Variable Rate Technology(VRT)
6. Precision irrigation systems
7. Remote controlled vehicle
8. Robot guidance
9. Combined with GIS

Automated machineries (tractors) have been developed that are operated with the help of Navigation Geographic Information System (NGIS) which is a combination of both GPS and GIS. Such facility enables the machine to display map, plan the path, control navigation, conduct analysis through sensor system, précising the positioning and communicate data to the end user. However, the user has a control over its speed, direction and can monitor the surrounding situations. The positioning information of the machines fed to them by GPS receiver that contains precise time, latitudes and longitudes that helps the machine by guiding over an optimal path. Such machines can also conduct farm activities like weeding and spraying. Thus, with the use of geospatial technologies (RS, GPS and GIS), farmers will be able to understand site specific needs of their farms according to which they will take decisions on management techniques, ensuring optimal use of inputs to maximize their outputs and profits.

Biotechnology

Biotechnology has evolved a lot over the past two decades in the field of agriculture showing its potential through "omics" transformation. One of the most successful result of such transformation are the genetically modified organisms (GMOs) /crops. Some benefits of genetic engineering in agriculture are increased crop yields, reduced costs for food production, reduced need for pesticides, enhanced nutrient composition and food quality, resistance to pests and disease, greater food security, and medical benefits to the world's growing population. Advances have also been made in developing crops that mature faster and tolerate aluminium, boron, salt, drought, frost, and other environmental stressors, allowing plants to grow in conditions where they might not otherwise flourish.

Apart from DNA related modifications as seen in GMOs, biotechnology has also paved its way through RNA interference (RNAi) technique to benefit agriculture, especially against their pests. This is a gene silencing mechanism in which the pests consume food (transgenic plant), in taking the double stranded RNA (dsRNA) which degrades the complementary mRNAi and inhibiting mRNA translation in pests, ultimately leading to its death or affecting its viability. In the past decade, the CRISPR–Cas9 technology has transformed genome engineering by removing the need for any expertise in engineering custom targeted DNA-binding proteins because the target specificity of CRISPR–Cas9 relies on base pairing of nucleic acids rather than protein–DNA recognition. It is a simple yet powerful tool for editing genomes. It allows researchers to easily alter DNA sequences and modify gene function. Its many potential applications include correcting genetic defects, treating and preventing the spread of diseases and improving crops.

Other contributions of biotechnology to agriculture:

1. Micro-propagation
2. Fortification of crops
3. DNA markers in aquaculture
4. Breeding and reproducing in aquaculture
5. GMOs for herbicide tolerance
6. Production of bio-fuel from bio-waste
7. Diagnostics of pests (ELISA, PCR, rapid test kits)

Nanotechnology

It is an emergent field of research that holds vivid applications in agriculture sector and food industry. In agriculture, nano-fertilizers are helping to enhance plant growth by providing smart nourishment. Nano-pesticides (control pests), nano-insecticides (against insects), nano-fungicides, and nano-herbicides (control weeds in the field) are the output of advanced nano-technological research. Nanoparticles are also used in seed science to enhance seed germination. The technological breakthrough in nanotechnology was achieved after the development of nano-sensors. Nano-biosensors are available for detecting phytopathogens and pesticide residues. In food sectors, nano particles are utilized in food processing and packaging. Besides this, nano-additives, nutraceutical delivery agents, and nano-encapsulation devices find multiple applications in several fields. Nano particles have been effectively used in the food packaging to enhance storage period and decrease pathogenic growth on stored food. Nano-biosensors are utilized to detect the contamination of food borne pathogens, toxic substance, and pesticide detection in food. Despite a lot of advantages of nanotechnology, the environmental toxicity of nano particles is a matter of concern that requires risk assessment of nano particles on human health and environmental effects.

The agro-nanotech innovative products are experiencing difficulties in reaching the market, making agriculture still a marginal sector for nanotechnology. This is due in particular to the high production costs of nanotech products, which are required in high volumes in the agricultural sector, unclear technical benefits and legislative uncertainties, as well as public opinion. Nevertheless, the R&D landscape is very promising and the possibilities offered by nanotechnology in several agricultural applications are being actively explored.

References

Digital Agriculture. (2018). Published by: DLG e.V., Eschborner Landstrasse, Frankfurt am Main

Grand View Research.(2019).Precision Farming Market Analysis. Estimates and Trend Analysis; Grand View Research Inc.: San Francisco, CA, USA,pp:1–58.

Kim, Y.J., Evans, R. G. and Iversen, W. M. (2008). Remote sensing and control of an irrigation system using a distributed wireless sensor network, Trans. Instrum. Meas., 57 (7): 1379-1387

L. Manivannan and M.S. Priyadharshini. (2016). Agricultural robot. International Journal of Advanced Research in Electrical, Electronics and Instrumentation Engineering, pp. 153-156

M. Dursun and S. Ozden. (2011). A wireless application of drip irrigation automation supported by soil moisture sensors, Sci. Res. Essays, 6 (7): 1573-1582.

Sarni, W., Mariani, J. and Kaji, J. From Dirt to Data: The Second Green Revolution and IoT. Deloitte

Zhang, Y. (2019). The Role of Precision Agriculture. Resource

2

Characterization of Chickpea Genotypes Based on Morphological Traits

L.N. Gediya, D.A. Patel, D. Kumar, K.J. Vekariya and J.J. Nayak

Introduction

Chickpea (*Cicer arietinum* L.) popularly known as Gram, Bengal gram, Egyptian pea, Chana, or Kabuli chana, Garbanzo bean is one of the first grain legumes to be domesticated by humans in old world (Van der Maesen, 1972). The genus *Cicer* belongs to the sub-family Papilionaceae of the family Leguminoceae (Bentham and Hooker, 1970) now popularly known as '*Fabaceae*' and have chromosome number $2n = 2x = 16$. The origin of the crop is considered to be Western Asia from where it has been transferred in India and other parts of the world (Rathore and Sharma, 2003). Importance of pulses is relatively more in our country as its contribution in nutrient supply is far more in Indian diet than that in Asia and world as a whole (Ali,2002).

Generally the quality of seed is estimated by varietal purity including physical and genetical. A variety/cultivar is an assemblage of cultivated plant which is clearly distinguished by any character (morphological, physiological, cytological, chemical or others) and which when reproduced (sexually or asexually) retains its distinguishing characters. Practically, a variety must show, Distinct, Uniform and Stable (DUS) variations in the characters that are adopted for use in varietal identification. To achieve this mission seed certification schemes have been launched to ensure cultivars identity and purity in the market place. Certification of identity has relied primarily on morphology (Singh, 2001).

Emphasis on characterization, varietal identification and genetic purity assessment of chickpea cultivars is very important to the field functionaries, certification officers, seed production officers and seed growers for regulating quality of the seed. However, standardized procedures are not available to seed analysts for determining cultivar purity. As a general rule, no cultivar can be identified or rejected purely by examining only seed or morphological characters in the field. Therefore, for keeping the purity of cultivars, stable visual diagnostic characters of seed, seedling and plant morphology are utmost essential to know (Lalitha, 2007).

Materials and Methods

Present investigation was carried out at Genetics and Plant Breeding Farm, B. A. College of Agriculture, A.A.U., Anand during Rabi of 2016-17. The experimental material comprised of 58 diverse chickpea genotypes including desi and kabuli types obtained from different chickpea research station of India. Different genotypes were evaluated in three replications using Randomized Complete Block Design (RCBD) with each entry in a single row of 4.0 m length (40 plants/genotype) in each replication. Inter and intra-row space was 30 × 10 cm, respectively. The experiment was surrounded by guard row to avoid damage and border effect. The recommended agronomical and plant protection practices were followed for the successful raising of the crop. The observations were recorded on 10 randomly selected plants for each character in each replication at different crop growth stages. Data were collected on 18 qualitative traits including morphological and seed characters.

Results and Discussion

Based on variation in physical characteristics, it was attempted to group the chickpea genotypes and identify each and every one of them through descriptors. Based on morphological variation, the 58 genotypes could be identified from each other. Based on anthocyanin pigmentation, all the 58 genotypes were grouped into two groups, with 34 genotypes showing anthocyanin pigmentation and the remaining 24 had absence of anthocyanin pigmentation (Figure 1 and Table 1). Four genotypes showed low insertion of first flower (<8 nodes), three genotypes showed medium insertion (8-15 nodes) and remaining 51 genotypes showed high insertion of first flower(>15 nodes). Variation were observed among genotypes for time of 50% flowering and based on this, genotypes were classified into three group. Thirty three genotypes were early with 40-60 days to 50% flowering, while 22 genotypes were medium with 61-80 days to 50% flowering and only three genotypes showed extra early flowering(<40 days). No variation was found among the genotypes for number of flowers per peduncle and all genotypes had single flower per peduncle. Plant growth habit is a distinguishing feature in varietal characterization. Out of 58 genotypes, 47 genotypes were observed as erect type, 9 genotypes were semi-erect and the remaining two genotypes were spreading type of growth habit.

Table 1: Distinguishable morphological characters in chickpea genotypes:

(a) Characterization of Plants

	Genotypes	Anthocyanin pigmentation	Height of insertion of first flower	Time of 50% flowering	Growth habit	Color of foliage	Leaflet size
1	GG-2	Present	High	Early	Erect	Medium green	Large
2	Vishal	Present	High	Early	Erect	Medium green	Small
3	JG-130	Present	High	Early	Erect	Medium green	Medium
4	GNG-14	Present	High	Medium	Erect	Medium green	Medium
5	GNG-27	Present	High	Medium	Erect	Medium green	Medium
6	GNG-28	Present	High	Early	Erect	Medium green	Medium
7	ICC-157	Present	Medium	Medium	Erect	Dark green	Small
8	ICC-231	Present	Medium	Medium	Erect	Dark green	Small
9	ICC-1164	Absent	High	Medium	Erect	Dark green	Small
10	ICC-1431	Present	High	Medium	Erect	Dark green	Small
11	ICC-2210	Present	High	Medium	Erect	Dark green	Medium
12	ICC-2277	Absent	High	Medium	Semi spreading	Light green	Small
13	ICC-2919	Absent	High	Early	Erect	Light green	Medium
14	ICC-3230	Present	High	Early	Erect	Dark green	Medium
15	ICC-4463	Present	High	Early	Semi spreading	Medium green	Small
16	ICC-4639	Present	High	Medium	Erect	Medium green	Small
17	ICC-4872	Absent	Low	Early	Semi spreading	Light green	Small
18	ICC-5037	Absent	Low	Medium	Erect	Dark green	Small
19	ICC-7255	Absent	High	Early	Semi spreading	Medium green	Large
20	ICC-8058	Absent	High	Medium	Erect	Medium green	Medium
21	ICC-8151	Absent	High	Medium	Semi spreading	Light green	Large
22	ICC-8195	Present	High	Medium	Erect	Dark green	Small
23	ICC-8318	Absent	High	Early	Semi spreading	Light green	Small
24	ICC-8350	Present	High	Medium	Erect	Medium green	Medium

	Genotypes	Anthocyanin pigmentation	Height of insertion of first flower	Time of 50% flowering	Growth habit	Color of foliage	Leaflet size
25	ICC-8522	Present	Low	Early	Erect	Light green	Small
26	ICC-8621	Absent	High	Medium	Erect	Medium green	Medium
27	ICC-8855	Absent	High	Early	Erect	Light green	Medium
28	ICC-9942	Present	Low	Medium	Erect	Dark green	Small
29	ICC-10393	Present	High	Extra early	Erect	Greenish purple	Medium
30	ICC-10399	Present	High	Early	Erect	Medium green	Small
31	ICC-10945	Present	High	Early	Erect	Medium green	Small
32	ICC-12028	Absent	High	Medium	Erect	Light green	Small
33	ICC-12155	Absent	High	Early	Erect	Medium green	Medium
34	ICC-12328	Absent	High	Medium	Spreading	Light green	Medium
35	ICC-12492	Absent	High	Early	Erect	Medium green	Medium
36	ICC-12866	Present	High	Early	Erect	Purple greenish	Medium
37	ICC-12928	Present	High	Medium	Erect	Dark green	Small
38	ICC-13124	Present	High	Early	Semi spreading	Medium green	Medium
39	ICC-13219	Present	High	Early	Erect	Medium green	Medium
40	ICC-13283	Absent	High	Early	Erect	Medium green	Medium
41	ICC-13357	Absent	High	Medium	Semi spreading	Medium green	Medium
42	ICC-13441	Absent	High	Early	Erect	Medium green	Small
43	ICC-13524	Present	High	Medium	Erect	Light green	Small
44	ICC-13764	Absent	High	Early	Erect	Medium green	Small
45	ICC-14669	Present	High	Early	Erect	Medium green	Small
46	ICC-14799	Present	Medium	Medium	Erect	Medium green	Small
47	ICC-14831	Present	High	Early	Erect	Medium green	Small
48	ICC-15264	Absent	High	Early	Semi spreading	Light green	Small
49	ICC-15333	Absent	High	Early	Spreading	Light green	Small
50	ICC-15888	Absent	High	Early	Erect	Dark green	Small

	Genotypes	Anthocyanin pigmentation	Height of insertion of first flower	Time of 50% flowering	Growth habit	Color of foliage	Leaflet size
51	GG-1	Present	High	Early	Erect	Dark green	Medium
52	GJG-3	Present	High	Extra early	Erect	Medium green	Large
53	GG-5	Present	High	Early	Erect	Dark green	Medium
54	GJG-6	Present	High	Early	Erect	Dark green	Large
55	JG-16	Absent	High	Early	Erect	Medium green	Small
56	Dahod yellow	Absent	High	Early	Erect	Light green	Small
57	Digvijay	Present	High	Extra early	Erect	Greenish purple	Medium
58	Vijay	Present	High	Early	Erect	Light green	Small

Fig. 1: Stem: Anthocyanin pigmentation

Four groups were made based on intensity of green colour of foliage. Out of 58 genotypes, 14 genotypes showed light green colour, 28 genotypes showed medium, 14 genotypes showed dark green and the remaining two genotypes showed greenish purple colour of foliage. With regard to size of leaflets, the variation was observed and the genotypes were categorized into three main groups namely, small, medium and large. Thirty genotypes recorded small size (10 mm), 23 genotypes recorded medium (10- 15 mm) and the remaining five genotypes were had large (>15 mm) sized leaf lets (Figure 2). Three groups were made based on variation in flower colour. Flowers of 16 genotypes were white, 40 genotypes showed pink and the remaining two genotypes showed blue colour of flowers (Figure 3). Genotypes were examined for the stripes on standard petal of the flower. In 19 genotypes, the stripes were absent and the remaining 39 genotypes showed stripes on standard of petal (Figure 3). The study of plant peduncle length revealed that chickpea genotypes can be classified into three categories. These even genotypes had short peduncle, 22genotypesrecordedas medium and the remaining 29 genotypes showed long peduncle (Figure 4). Based on plant height all genotypes were classified in to three groups. Five genotypes were short in plant height (<45 cm), 27 genotypes recorded medium plant height (45-65 cm) and the remaining 26 genotypes were tall in plant height (>65 cm) (Figure 5). Size of the pod was used in characterizing the chickpea genotypes. Based on pod size, genotypes were classified into three groups. Twelve genotypes were recorded as small (<15 mm), 32 genotypes recorded medium (15-20 mm) and the remaining 14 genotypes were classified as large (>20 mm) sized pods (Figure 6). Difference were also found in number of seeds per pod, 29 genotypes showed one seed per pod and remaining 29 genotypes showed two seeds per pod.

Small Medium Large

Fig. 2: Leaflet: size (length) (Middle of the plant and middle of the leaf)

Pink & Present Whiite & Absent Blue & Present

Fig. 3: Flower: Colour and Stripes on standard of flower

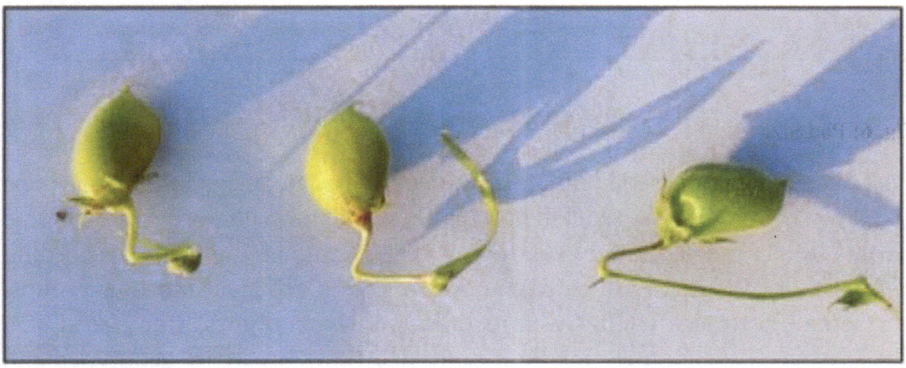

Small Medium Long

Fig. 4: Peduncle Length

Short Medium Tall

Fig. 5: Plant Heigh

Small Medium Large

Fig. 6: Pod Size

Based on variation in seed colour, genotypes were categorized into ten groups viz., brown, dark brown, reddish brown, beige (kabuli with pure white), black, creamy beige, green, orange, yellow and grey in colour (Figure 7). According to seed size (weight of 100 seed), genotypes were grouped into four groups. Very small seed size (<20g) was recorded in 32 genotypes, small seed size (20-25g) in 10 genotypes, medium seed size (26-35g) in 14 genotypes and large seed size (36-45g) was recorded in remaining three genotypes (Figure 8). Significant differences were found among genotypes for seed shape and based on this, the genotypes were classified in to three groups. Eight genotypes had pea shape, 29 genotypes had owl's head and the remaining 21 genotypes had angular seed shape (Figure 9). Three groups were made based on seed testa texture. Rough texture was observed in 28 genotypes, smooth texture in 23 genotypes and tuberculated in remaining 7 genotypes (Figure 10). According to seed ribbimg, genotypes were categorized into two groups. Forty two genotypes showed presence of seed ribbing and remaining 16 genotypes showed absence of seed ribbing (Figure 11). Among the 58 genotypes, 46 genotypes were desi and remaining 12 genotypes were kabuli type (Figure 12). Morphological features of genotypes have been a major component of varietal identification. It is not possible to identify varieties using any single parameter. A detailed morphological description of plants and seeds should therefore be prepared. Utilization of these features in sequential fashion is useful and convenient to distinguish different genotypes.

Fig. 7: Seed: Colour (one month after harvest using colour chart)

Fig. 8: Seed Size (weight of 100 seeds at 10% moisture content)

Pea Shaped Owl's head Angular

Fig. 9: Seed Shape

Rough Smooth Tuberculate

Fig. 10: Seed: Testa texture

Absent Present

Fig. 11: Seed: Ribbing

Desi Kabuli

Fig. 12: Seed Type

The local type are characterized by small, irregularly shaped seeds of various colours (light brown, dark brown, yellow pink, green, black or variegated colours) having thick and rough seed coat with pinkish or bluish flowers, small plant stature small leaflets, presence of anthocyanin pigmentation and usually 2-3 seeds per pod. The kabuli type have rams head or round or large owl's head shape, cream colour, larger seed with smooth seed coat, long plant stature, white flowers, bigger leaflets, no anthocyanin and usually one or two seeds per pod. There is no basic difference in chromosome number (2n=16) of local and kabuli and both types cross each other readily.

Similarly, genotypes identification based on distinguishable morphological characters were reported by Lalitha (2007), Upadhyaya *et al.* (2002), Yadav and Shrivastava (2002), Chowdhury *et al.* (2002) and Singh (2001) in chickpea.

Table 1 (b): Characterization of Flower

	Genotypes	Flower color	Stripes on standard of flower	Peduncle length	Plant height	Pod size	No. of seeds per pod
1	GG-2	Pink	Absent	Medium	Medium	Medium	One
2	Vishal	Pink	Present	Long	Medium	Large	One
3	JG-130	Pink	Present	Long	Medium	Medium	One
4	GNG-14	Pink	Present	Medium	Medium	Medium	One
5	GNG-27	Pink	Present	Medium	Medium	Large	Two
6	GNG-28	Pink	Present	Long	Medium	Medium	Two
7	ICC-157	Pink	Present	Medium	Short	Medium	Two
8	ICC-231	Pink	Present	Long	Short	Small	One
9	ICC-1164	White	Absent	Long	Medium	Medium	Two

	Genotypes	Flower color	Stripes on standard of flower	Peduncle length	Plant height	Pod size	No. of seeds per pod
10	ICC-1431	Pink	Present	Medium	Medium	Medium	One
11	ICC-2210	Pink	Present	Medium	Medium	Small	One
12	ICC-2277	White	Absent	Medium	Tall	Large	One
13	ICC-2919	Pink	Present	Medium	Tall	Medium	One
14	ICC-3230	Pink	Present	Long	Medium	Small	One
15	ICC-4463	Pink	Present	Medium	Tall	Medium	Two
16	ICC-4639	Pink	Present	Medium	Medium	Medium	Two
17	ICC-4872	Pink	Present	Long	Short	Medium	One
18	ICC-5037	White	Absent	Medium	Medium	Small	One
19	ICC-7255	White	Absent	Long	Tall	Large	Two
20	ICC-8058	White	Absent	Medium	Tall	Large	Two
21	ICC-8151	White	Absent	Medium	Tall	Large	One
22	ICC-8195	Pink	Present	Medium	Medium	Small	One
23	ICC-8318	Pink	Present	Long	Medium	Medium	One
24	ICC-8350	Pink	Absent	Long	Tall	Large	One
25	ICC-8522	Pink	Present	Long	Medium	Small	Two
26	ICC-8621	Pink	Present	Medium	Tall	Medium	Two
27	ICC-8855	White	Absent	Medium	Tall	Large	Two
28	ICC-9942	Pink	Present	Medium	Short	Small	One
29	ICC-10393	Pink	Present	Long	Medium	Medium	Two
30	ICC-10399	Pink	Present	Medium	Medium	Medium	Two
31	ICC-10945	Pink	Present	Long	Medium	Medium	Two
32	ICC-12028	Pink	Present	Medium	Tall	Medium	One
33	ICC-12155	Pink	Present	Short	Medium	Medium	Two
34	ICC-12328	White	Absent	Long	Tall	Large	One
35	ICC-12492	White	Absent	Short	Medium	Medium	Two
36	ICC-12866	Pink	Present	Long	Tall	Small	Two
37	ICC-12928	Pink	Present	Short	Tall	Small	One
38	ICC-13124	Pink	Present	Long	Tall	Medium	One
39	ICC-13219	Pink	Present	Medium	Medium	Medium	Two
40	ICC-13283	White	Absent	Short	Tall	Large	One
41	ICC-13357	White	Absent	Short	Tall	Medium	One
42	ICC-13441	White	Absent	Long	Medium	Medium	One
43	ICC-13524	Pink	Present	Short	Tall	Small	Two

	Genotypes	Flower color	Stripes on standard of flower	Peduncle length	Plant height	Pod size	No. of seeds per pod
44	ICC-13764	White	Absent	Long	Tall	Medium	Two
45	ICC-14669	Pink	Present	Long	Medium	Medium	One
46	ICC-14799	Pink	Present	Medium	Short	Small	One
47	ICC-14831	Pink	Present	Long	Tall	Medium	Two
48	ICC-15264	White	Absent	Medium	Tall	Medium	Two
49	ICC-15333	White	Absent	Long	Tall	Large	One
50	ICC-15888	White	Absent	Long	Medium	Small	Two
51	GG-1	Pink	Present	Long	Medium	Medium	Two
52	GJG-3	Pink	Present	Long	Tall	Large	Two
53	GG-5	Pink	Absent	Long	Tall	Medium	Two
54	GJG-6	Pink	Present	Long	Medium	Large	One
55	JG-16	Pink	Present	Long	Medium	Medium	Two
56	Dahod yellow	Pink	Present	Short	Tall	Medium	Two
57	Digvijay	Pink	Present	Long	Tall	Large	Two
58	Vijay	Pink	Present	Long	Tall	Medium	One

Table 1: © Characterization of Seed

S.No.	Genotypes	Seed color	Seed size	Seed shape	Seed testa texture	Seed ribbing	Seed type
1	GG-2	Brown	Medium	Owl's head	Rough	Present	Desi
2	Vishal	Dark brown	Medium	Owl's head	Rough	Present	Desi
3	JG-130	Brown	Medium	Angular	Rough	Present	Desi
4	GNG-14	Brown	Medium	Pea shaped	Rough	Absent	Desi
5	GNG-27	Green	Very small	Owl's head	Rough	Present	Desi
6	GNG-28	Brown	Very small	Angular	Tuberculate	Present	Desi
7	ICC-157	Brown	Very small	Angular	Rough	Present	Desi
8	ICC-231	Brown	Very small	Owl's head	Smooth	Present	Desi
9	ICC-1164	Orange	Very small	Pea shaped	Rough	Absent	Desi
10	ICC-1431	Brown	Small	Owl's head	Smooth	Absent	Desi
11	ICC-2210	Yellow	Very small	Owl's head	Smooth	Present	Desi
12	ICC-2277	Creamy beige	Medium	Owl's head	Rough	Absent	Kabuli
13	ICC-2919	Beige Kabuli	Small	Angular	Rough	Present	Kabuli
14	ICC-3230	Green	Very small	Owl's head	Rough	Present	Desi
15	ICC-4463	Black	Very small	Angular	Smooth	Present	Desi
16	ICC-4639	Brown	Very small	Angular	Rough	Present	Desi
17	ICC-4872	Orange	Small	Pea shaped	Rough	Absent	Desi

S.No.	Genotypes	Seed color	Seed size	Seed shape	Seed testa texture	Seed ribbing	Seed type
18	ICC-5037	Green	Small	Pea shaped	Smooth	Absent	Desi
19	ICC-7255	Creamy beige	Medium	Owl's head	Smooth	Present	Kabuli
20	ICC-8058	Creamy beige	Very small	Owl's head	Smooth	Absent	Desi
21	ICC-8151	Creamy beige	Large	Owl's head	Tuberculate	Present	Kabuli
22	ICC-8195	Grey	Very small	Owl's head	Smooth	Present	Desi
23	ICC-8318	Yellow	Very small	Angular	Tuberculate	Present	Desi
24	ICC-8350	Reddish brown	Medium	Pea shaped	Rough	Absent	Desi
25	ICC-8522	Black	Very small	Angular	Smooth	Present	Desi
26	ICC-8621	Creamy beige	Small	Owl's head	Rough	Present	Kabuli
27	ICC-8855	Creamy beige	Small	Pea shaped	Smooth	Absent	Desi
28	ICC-9942	Brown	Very small	Angular	Rough	Present	Desi
29	ICC-10393	Reddish brown	Very small	Angular	Smooth	Present	Desi
30	ICC-10399	Brown	Very small	Angular	Smooth	Present	Desi
31	ICC-10945	Dark brown	Very small	Angular	Tuberculate	Present	Desi
32	ICC-12028	Kabuli beige	Medium	Angular	Smooth	Present	Kabuli
33	ICC-12155	Brown	Very small	Angular	Rough	Present	Desi
34	ICC-12328	Creamy beige	Large	Owl's head	Smooth	Present	Kabuli
35	ICC-12492	Creamy beige	Very small	Pea shaped	Smooth	Absent	Desi
36	ICC-12866	Yellow	Very small	Angular	Rough	Present	Desi
37	ICC-12928	Yellow	Very small	Owl's head	Smooth	Present	Desi
38	ICC-13124	Reddish brown	Medium	Owl's head	Smooth	Present	Desi
39	ICC-13219	Yellow	Very small	Angular	Tuberculate	Present	Desi
40	ICC-13283	Creamy beige	Medium	Owl's head	Smooth	Absent	Kabuli
41	ICC-13357	Creamy beige	Medium	Owl's head	Rough	Present	Kabuli
42	ICC-13441	Creamy beige	Small	Owl's head	Rough	Absent	Kabuli
43	ICC-13524	Black	Very small	Angular	Smooth	Present	
44	ICC-13764	Creamy beige	Very small	Owl's head	Smooth	Absent	Desi

S.No.	Genotypes	Seed color	Seed size	Seed shape	Seed testa texture	Seed ribbing	Seed type
45	ICC-14669	Yellow	Very small	Angular	Tuberculate	Present	Desi
46	ICC-14799	Yellow	Very small	Angular	Rough	Present	Desi
47	ICC-14831	Yellow	Very small	Owl's head	Rough	Present	Desi
48	ICC-15264	Creamy beige	Medium	Owl's head	Smooth	Absent	Kabuli
49	ICC-15333	Creamy beige	Large	Pea shaped	Tuberculate	Absent	Kabuli
50	ICC-15888	Orange	Very small	Owl's head	Rough	Absent	Desi
51	GG-1	Dark brown	Very small	Owl's head	Smooth	Present	Desi
52	GJG-3	Brown	Medium	Owl's head	Rough	Present	Desi
53	GG-5	Dark brown	Small	Owl's head	Smooth	Present	Desi
54	GJG-6	Dark brown	Small	Owl's head	Rough	Present	Desi
55	JG-16	Brown	Very small	Owl's head	Rough	Present	Desi
56	Dahod yellow	Yellow	Very small	Angular	Rough	Present	Desi
57	Digvijay	Dark brown	Medium	Owl's head	Rough	Present	Desi
58	Vijay	Brown	Small	Angular	Rough	Present	Desi

References

Ali M. (2002). Pulses Production Scenario in India: Present status and future strategies. In booklet published on the occasion of short course on "Recent Advances in methods of improvement of pulse crops" at IIPR, Kanpur during 27th May – 5th June,

Bentham G, Hooker JP. (1970). Genera platinum (Genera of plants), Reeve & Co., London.; 1:324.

Chowdhury MA, Bert Vanden berg, Warkentin T. (2002) Cultivar identification and genetic relationship among selected breeding lines and cultivars in chickpea (*Cicer arietinum* L.). *Euphytica.*; 127(1-2):317-325. Guidelines for the Conduct of Test for Distinctness, Uniformity and Stability on Chickpea (*Cicer arietinum* L.). http://www.plantauthority.gov.in/pdf/GChickpea.pdf (Accessed August 04, 2018)

Kumar Y, Ram H, Gulia SK. Varietal identification in oat based on field parameters. Forage Res. 2002; 28(2):73-6.

Kumar S, Yadav TP, Ram C. Varietal identification through stable descriptors in rapeseed and mustard (Brassica sps. viz., Juncea campestris, *Napus* and *Eruca sativa*). Seed Tech News, 2000; 28(4):50.

Lalitha A. Varietal Characterization of chickpea (*Cicer arietinum*) varieties (Doctoral dissertation, Acharya N. G. Ranga Agricultural University, Rajendranagar, Hyderabad) 2007.

Rathore PS, Sharma SK. Scientific pulse production, Yash Publishing House, Bikaner, Rajasthan. 2003;92.

Singh R. Characterization of chickpea cultivars by field and laboratory techniques (Doctoral dissertation, Chaudhary Charan Singh Haryana Agricultural University, Hisar) 2001.

Upadhyaya HD, Ortiz R, Bramel PJ, Singh S. Phenotypic diversity for morphological and agronomic characteristics in chickpea core collection. Euphytica. 2002 Feb 1; 123(3):333- 42.

Van der Maesen LJG. Cicer L., A monograph of the genes with special reference to the chickpea, its ecology and cultivation, Biol. J. Linn. Soc. Wageningen, the Netherlands. 1972.

Yadav RDS, Shrivastava JP. DUS characteristics of chickpea varieties. Seed Tech News. 2002; 32(1), 29-30.

3

Green Synthesis and Characterization of Magnesium Oxide Nanoparticles for Inducing Drought Tolerance in Rice

S.K. Dwibedi, M. Mohanty, P. Shatpathy, A. Dash, M. Behera and R.K. Dwibedi

Introduction

Rice (*Oryza sativa* L.) is the vital cereal crop, probably the oldest cultivated grain crop, produced under varied ecosystems is consumed as the primary source of food by the billions of people throughout the world covering 9% of total arable land among world. India is one of the top leading countries in the entire world in case of both production and consumption of rice. Rice is cultivated in 43.86 m ha geographical area of the country producing 115.6 million tons (2018- 2019) of rice. It is produced in various ecosystems throughout the country under diverse climatic and soil condition but the productivity is much lower as compared to other countries. According to Pandey *et al.* (2006), the average loss of rice production during drought years in major rainfed rice grown states like Chhattisgarh, Jharkhand and Odisha was found to be 30 percent of the annual production in case of non-drought years. Drought being one of the major climate related abiotic stress has significant effect on rice plant leading several physiological, morphological changes in the plant bodies. Drought severely affects rice production by affecting more than 23 m ha of total rainfed rice area in Asia (Kumbhar *et al.*, 2015). It acts as a major limiting factor for crop production by affecting gene expression, crop growth, quality and stability of yield which causes severe problem in many parts of the world (Passioura, 2007). Aberrant weather conditions such as delayed onset of monsoon or prolonged dry spells lead to deficit soil moisture stress which limits crop yield, grain quality and gene expression. Plants adapt several physiological, biochemical, morpho-anatomical and molecular mechanisms to come over such water stress. Development of tolerant varieties of rice for deficit moisture stress is challenging, time consuming and costlier, hence many alternative strategies are being experimented to induce tolerance in rice to moisture deficit stress. Seed germination and early seedling growth have been reported as the critical stage for drought stress and according to many research reports, drought stress reduces the germination of seed, seedling height and no. of tillers in rice (Sokoto and Muhammad, 2014).

Nanotechnology is a versatile field that has found applications in the field of science. Application of nanoparticles in the plants for the purpose of controlling plant diseases and changing plant's physiochemical response is a practice of recent times. Nanoparticles are the ultrafine particles having size of 1-100 nanometers. Nanoparticle acts as magic bullet which contains genes, chemicals or herbicides which are released into plant parts after application in plants (Mukherjee *et al.*, 2014). The increased reactivity and biochemical activity of nanomaterials is attributed due to their high surface to volume ratio. These materials easily enter into plants by overcoming all barriers at cellular level than that of bulky material. Due to their physicochemical characteristics nanoparticles has the potential to modulate the redox status and affect the germination, growth, quality and performance of the plants (Adams, 2006). However, the interaction mechanism between the nanomaterial and the system is unknown. In agriculture, despite of being an important factor for growth and development of plants, most of the fertilizer after application becomes unavailable because of several factors such as degradation by photolysis, hydrolysis and degradation and leaching. Therefore, to enhance crop yield and reduce the nutrient loss from fertilizers, nanotechnology and nanomaterial application will be helpful.

Magnesium oxide nanoparticle has been a topic of recent research activity in nanotechnology. MgO nanoparticle has also been studied regarding its antibacterial properties against foodborne pathogens. Jayarambabu *et al.* (2016) studied the beneficial role of MgO nanoparticle on seed germination and plant growth regulation in maize by increasing the activity of growth hormones. The nanoparticle size also has an important role in growth parameters of maize. As MgO-NPs can easily enter plant cells, it affects chlorophyll content and plant biomass. However, many top down and bottom up methods have been standardized for synthesis of Mgo- Nps nanoparticles but such methods are costly and require costly equipments. In this experiment successful attempts were as made to synthesize MgO-NPs from red *Amaranthus* sps for inducing drought tolerance in rice during initial growth phases.

Materials and Methods

The Magnesium oxide (MgO) nanoparticles were synthesized in laboratory from red *Amaranthus* sps. leaves following green chemistry method. The synthesis was three stage processes which are described below.

Fresh red Amaranthus plants were purchased from local market and the leaves were separated from the stem. After thorough washing of 20 g of leaves with tap water, the samples were dried inside room for 20 minutes. The excess water was removed by blotting paper. The room dried fresh leaves were then macerated with distilled water by using mortar and pestle. The substrate was then transferred into 200 ml glass beaker and volume was made upto 200 ml with distilled water followed by heating in hot plate at 80 ^{0}C for 1 hour. The entire substrate was then filtered into another beaker by using Whatman's No.1 filter paper. 40 ml of this substrate was pipetted out into a 500 ml beaker with 70 ml distilled water and 5 g of Magnesium nitrate [Mg(NO3)2.6H2O] was added into it on a magnetic stirrer-cum-hot plate at 60 ^{0}C and 1,000 rpm for 1 hour. During stirring, 1% Ammonium hydroxide (NH4OH) solution was added into it drop wise to allow the reaction under alkaline medium. The initial reddish colour of the solution changed to dark brown indicating the formation of Magnesium hydroxide nanoparticles.

The solution was then centrifuged at 10,000 rpm in a centrifuge machine for 10 minutes and the substrates thus settling at the bottom were collected in a crucible and oven-dried for calcination. The calcination of sufficient quantities of these substrates was done in muffle furnace at 500 ^{0}C for 1.5 hours to get the grayish white ash coloured MgO nanoparticles. The nanoparticles were then stored at 4 ^{0}C in the laboratory for further characterization and application.

The MgO powder thus obtained was sent to the Central Institute of Plastics Engineering and Technology (CIPET), Bhubaneswar and Central Instrumentation Facility (CIF), OUAT, Bhubaneswar for characterization by Transmission Electron Microscope (TEM), Power X-Ray Diffraction (XRD), Ultraviolet–Visible Spectrophotometry/ Spectroscopy (UV-VIS), and Fourier-Transform Infrared Spectroscopy (FTIR) analyses.

Ultraviolet visible spectrophotometer: It is used to obtain the peak absorbance reading of the calcined powdery mass confirming the presence of Mg in it.

Transmission Electron Microscope (TEM): TEM uses an electron beam to measure size and distribution of nanoparticles, providing much higher resolution than with light-based imaging techniques

Fourier transform infrared spectroscopy (FTIR): By using Infrared light, it scans the test sample and expresses the chemical properties.

X-Ray diffraction (XRD): From the XRD graph, average crystallite size in the Mg O nanoparticle was calculated by following Debye-Scherrer formula as:

$$D = \frac{k\lambda}{\beta \cos\theta}$$

Results and Discussion

a) **UV/VIS Absorption Spectroscopy:** UV/VIS absorption spectroscopy is most widely used analytical technique for characterizing the electronic structure of the optical band gap of the nano material. The spectrum shows (Fig. 1) a strong band in the UV region due to the charge transfer transition between energy states. Firstly, the absorbance decreases sharply with increase in the wavelength near the band edge 210 nm that indicated the size of MgO-NP.

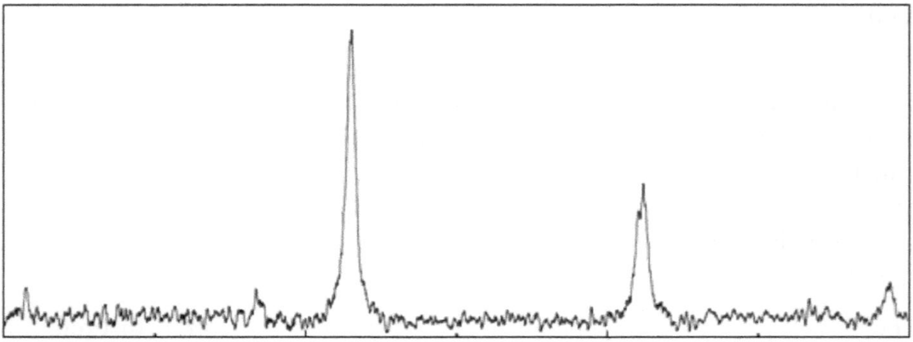

Fig. 1: UV/VIS absorption spectra of phytosynthesised Magnesium oxide sonicated in ethanol medium

b) **FTIR Spectra analysis:** FTIR spectra of the MgO-NP, calcined at 500 ^{0}C for 1 hr was recorded on a Perkin Elmer RXFTIR Spectrometer (Fig. 2). Perusal of the figure shows the IR broad peaks at around 3,296.85 cm^{-1}, 1,431.41 cm^{-1}, 1,046.72 cm^{-1}, 856.54 cm^{-1}and 673.23 cm^{-1}. A broad band between 3,600 to 3,200 cm^{-1} are due to stretching mode of alcohol -OH group, peak at 1431.41 cm-1 was due to –CH2 group, whereas at 1,046.72 cm-1 was due to primary alcohol. The source of alcohol could be ethanol which was used in laboratory for washing the $Mg(OH)^2$ during centrifuge process. The peaks at 856.54 and 673.23 cm^{-1} were attributed to different Mg-O-Mg vibration modes of MgO (Jayarambabu et al.,2016).

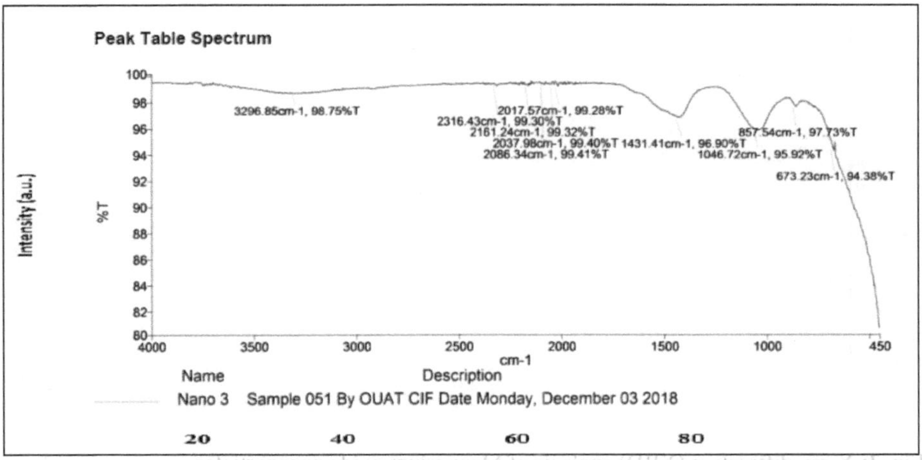

Fig. 2: FTIR spectra of MgO in phytosynthesised nano powder

c) **X-ray diffraction (XRD) analysis:** The XRD pattern of the MgO sample, calcined at 500 °C for 1 hr, is presented in Fig.3. All the corresponding peaks were indexed to cubic MgO crystallites with the lattice parameter a = 0.421 nm using the standard values of JCPDS 45-946. Peaks were observed at 36°, 42°, 62°, 74° and 78° corresponding to miller indices values (111), (200), (220), (311) and (222), respectively. No peak arising from an impurity was observed in Figure 3. The average particle size in the calcined MgO nanoparticles was estimated by the Scherrer equation.

$$D = \frac{k\lambda}{\beta \cos\theta}$$

Where, D is the crystallite size, k is the shape factor, which usually takes a value of about 0.89, λ is the wavelength of X-ray source used (0.15406 nm for Cu Kα), β is the full width at half-maximum (FWHM) in radians (0.1117), and θ is the Bragg diffraction angle in radians (0.374558). The average particle size of MgO, calcined at 500 °C, was measured to be 13.2 nm.

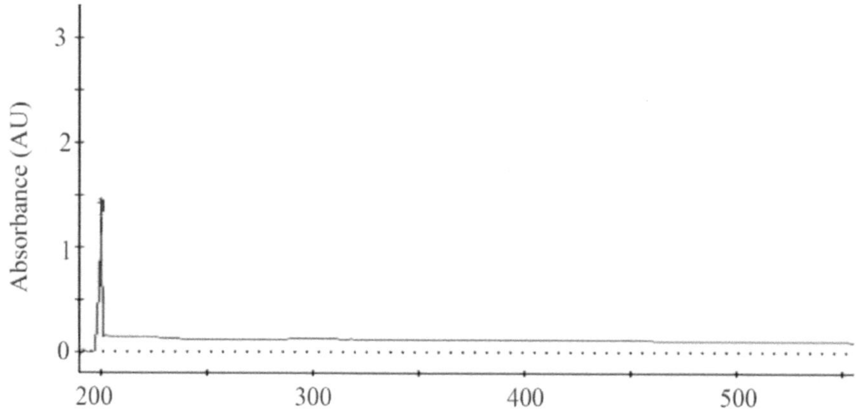

Fig. 3: X-ray diffraction (XRD) analysis of Magnesium oxide nanoparticles

d) Transmission Electron Microscope (TEM)analysis: The Fig. 4 and 5 illustrate TEM image of MgO-NP. From the image it is observed that a small amount of agglomeration present in the as-synthesized sample. The MgO particles are nearly monodisperse and the average crystallite size seen in the micrograph is of the order of 10 nm. Both XRD and TEM analyses of samples are in concurrence with each other. The crystal lattice space (gap between two parallel red lines in Fig. 5 is 0.21 nm which is in coherence with the XRD analysis report with miller index 200.

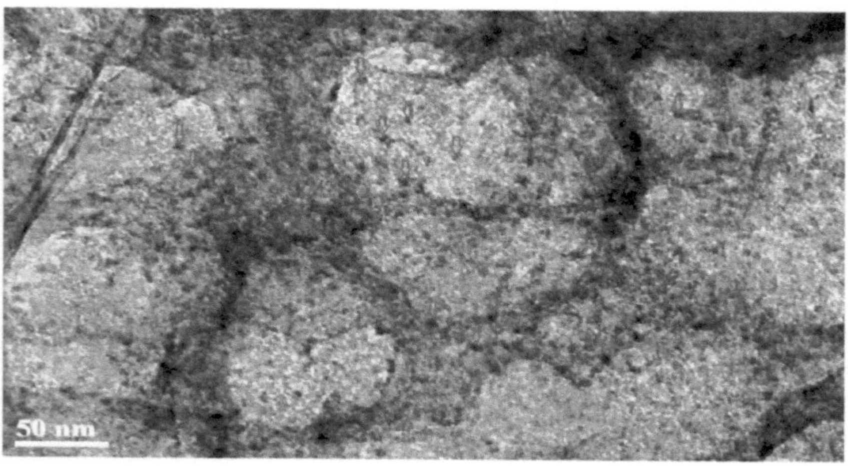

Fig. 4: Magnesium oxide nanoparticles (red arrow marks) studded in the amorphous ash

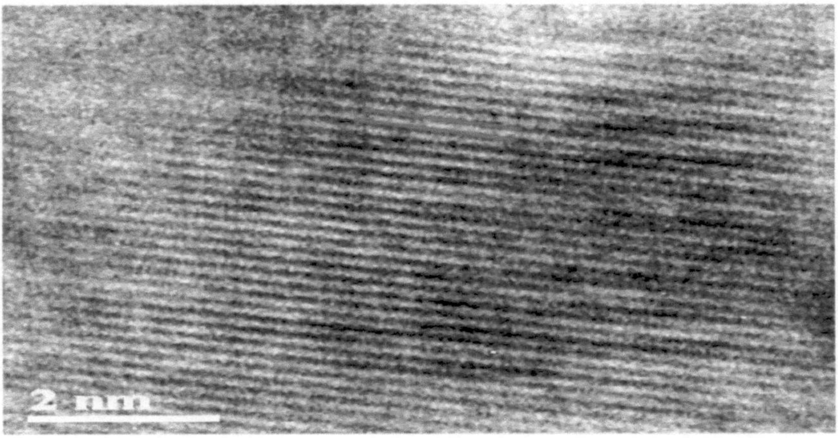

Fig. 5: High resolution (filtered) TEM image of Magnesium oxide nanoparticles. Lattice spacing are measured as 0.21054 nm corresponding to (200) crystal plane

Conclusion

The MgO nanoparticles synthesized by green chemical bottom up method were of crystalline structure with average size of 13.2 nm as characterized by UV/VIS, FTIR, XRD and TEM. The MgO-NPs thus synthesized were used for inducing drought tolerance in rice during germination and seedling stage.

Acknowledgement

The technical and informative supports of the Central Instrumentation Facility (CIF), OUAT; Laboratory for Advanced Research of Polymeric Materials (LARPM), Central Institute of Plastics Engineering & Technology (CIPET), Bhubaneswar; and Institute of Physics (IOP), Bhubaneswar are highly acknowledged.

References

Adams LK, Delina L, Mclntosh A and Alvareez PJJ. (2006). Comparative toxicity of nano-scale TiO_2, SiO_2 and ZnO water suspensions, Water Science & Technology, 54(11-12): 327-34.

Jayarambabu N, Siva Kumari B, Venkateswara RK and Prabhu YT. (2016). Enhancementof growth in maize by biogenic- synthesized MgO nanoparticles. International Journal of Pure and Applied Zoology, 4 (3): 262-270.

Kumbhar S, Kulwal P, Patil JV, Sarawate CD, Gaikwad AP, Jadhav AS.2015.Genetic Diversity and Population Structure in Landraces and Improved Varieties of Rice from India, Rice Science, 22(3):99-107.

Mukherjee A, Peralta-Videa JR, Bandyopadhyay S, Rico CM, Zhao L, Gardea-Torresdey JL. 2014. Physiological effects of nanoparticulate ZnO in green peas (*Pisum sativum* L.) cultivate in soil, Metallomics, 6(1):132-8.

Pandey S, Bhandari H, Ding S, Prapertchob P, Sharan R, Naik D, Taunk SK and Sastri A. (2006). Coping with Drought. In: Rice Farming in Asia: Insights from a Cross-Country Comparative Study.

Passioura, J. (2007). The drought environment: physical, biological and agricultural perspectives, Journal of Experimental Botany; 58(2): 113-117.

Sokoto MB and Muhammad A. (2014). Response of rice varieties to water stress in Sokoto, Sudan Savannah, Nigeria, Journal of Biosciences and Medicines, 2(1): 68–74.

4

Performance of Continuous and Pulse Drip Irrigation on Yield of Tomato (*Lycopersicon esculentum*)

D. Dash, S.C. Senapati, A.P. Sahu and P.K. Paramaguru

Introduction

Irrigated agriculture plays a predominant role in the state's food grain production. Erratic trend of rainfall, it has been realized that the availability of water will become very less in forthcoming days as a result cultivation of crops will face water crisis. Development of new water resources needs excessive money investment with high level of expertise and technical knowhow. The ground water is also falling down to greater and greater depth due to overdraft causing for the restriction of the construction of tube wells. Further the efficiency of traditional irrigation system in India is now varying from 30% to 40% against 62 to 65% during the project plan periodbecause of wastage of this valuable water resource due to mismanagement. To meet the increasing demand for additional food grain production, the demand for water in the irrigation sector will increase substantially. In such a situation conservation of water and its proper management is necessary for developing a sustainable agricultural system avoiding the degradation of the land quality, water and other natural resources in an irrigated area. At this juncture, drip / trickle irrigation is now in vogue is the frequent, slow application of water either directly on to the land surface or into the root zone of crop. Irrigating a portion of land surface limits the evaporation, reduces weed growth and minimizes interruption of cultural operation. Because of its higher water application frequency, concentration of salts remains manageable in the rooting zone (Mantell *et al.,* 1985).

Continuous water application through drip irrigation is associated with increased water percolation under root zone. Intermittent irrigation strategies based on discharge pulses followed by break will improve water management in the field and increase irrigation efficiency (Oron, 1981). Pulsing drip irrigation is a recent concept of irrigation method refers to the practice of irrigating for a short period and waiting for another short period and repeating this on-off cycle until the entire irrigation water is applied (Eric *et al.*, 2004). The higher the irrigation frequency, the smaller the soil wetted volume and the higher soil water content in a small range can be maintained. Other advantages of pulsing are that plant growth is generally greater than with standard irrigation and lower fertilizer rates can be used (Dole, 1994 and El-Mogy *et al,* 2012).

Pulse drip irrigation technique increased crop yield, saved water and increased water use efficiency and wetted soil volume (more than or equal to 100% of field capacity) in root zone (Zin El-Abedin, 2006; Bakeer *et al.,* 2009), slightly reduces the water losses under the root zone in cases with and without hysteresis (Elmaloglou and Diamantopoulos, 2008) as compared to continuous drip under surface drip irrigation.

The growth and quality of potato under organic agriculture with dripper discharge of 2 lph were improved by increasing irrigation frequency under 100% and 75% from water requirements and the same parameters were decreased at 50% from water requirements (El Saidi *et al.,* 2010).

Increasing the number of pulses for each irrigation in trickle irrigation system improved the vegetable growth and increasing the green bean yield from 3.92 t/fed (WUE 4.59 kg/m^3) in one time water application to 4.78 t/fed (WUE 6.84 kg/m^3) when water was applied in four pulses (El-Mogy *et al.* , 2012).

Clogging of emitters is one of the important parameter which affects the uniformity of water distribution. The emitter clogging can be minimized by increasing the emitter orifice size (Mushari, 2008; Al-Amoud and Saeed, 1988), increasing the size of the waterway in the emitters (Jackson *et al.,* 1987) and by increasing the number of irrigation pulses (Abdelraouf *et al.,*2012). Further to maintain high distribution uniformity and reduce clogging, high water application rates are advisable for shallow rooted and widely spaced crops where as lower application rates recommended for deep rooted and closely spaced row crops (Mushari, 2008).

The experiment was conducted in the farmers' field of village Khandanali in the district Angul of state Odisha (India) with tomato as test crop. Tomato is the second most important vegetable crop next to potato used in every day's life. It is highly nutritive, remunerative and has great processing potentiality. Canal is the major irrigation source in Angul district. Owing to erratic nature of rainfall and non availability of required quantity of water from canal system, the farmers of the district started opening of open dug wells from the upper zone of the earths' crust and few farmers have constructed deep wells for irrigation to *rabi* vegetable crops. At present, the ground water level also is going down ward due to over draft. There was no use of micro irrigation system. After 2011 very few farmers have adopted drip irrigation system. So there was requirement of creating awareness among the farmers by practical demonstration of drip irrigation system through research methodology in the field. The present study was undertaken to study the spatial variation of soil moisture, determination of application efficiency, delineating deep percolation losses and economics in cultivation of tomato crop with continuous and pulse drip irrigation.

Materials and Methods

Study area

The experiment was conducted in the farmers' field of village Khandanali about 15 kms away from the district headquarter Angul falling under mid central table agro-climatic zone of Odisha (India). The site is located at an altitude of about 440 m above mean sea level with a latitude of 20^0 85' N and Longitude is $85^0$10' E . The soil of the study area comprises of sand 81.2%, silt 5.4% and clay 13.4% and texturally classified as sandy loam. The bulk density, field capacity, wilting point, pH, E_c and organic content of the soil were tested in the laboratory and found to be 1.53 gm/cm^3, 13.2 %, 4.8 %, 5.8, 1.2 ds m^{-1} and 0.62 g kg^{-1} respectively. After harvest of paddy, most of the area is covered under vegetable and spices crops like tomato, brinjal, okra, cauliflower, coriander, pea, cabbage, chilli, onion, and garlic. Tomato is the most important crop after chilli. The productivity of tomato is 13.45t ha^{-1}(Anonymous, 2008-09) which is less than the national average of 17.50 t ha^{-1}. This is due to poor management practice of canal and underground water. Using ground water for farming, which is resulting steady depletion of ground water resources. So this study was taken with an aim to recommend new operational approaches of hi-tech vegetable production with optimal use of water.

Water from a bore well was used for irrigation to tomato crop through drip irrigation system. Water and soil analysis kit was used to measure the pH, EC and turbidity of the water in the tube well were found to be 5.73, 1.3 ds/m and 1 N.T.U. respectively. The EC and turbidity values indicate that water can be safely used for irrigation to tomato crop through drip system though the water is slightly acidic in nature.

Experimental design and nutrient management

The details of the treatments of the present work to irrigate tomato as test crop (cv. Chiranjibi Hyb.) through drip irrigation are given in Table 1. The experimental field was divided into three large plots representing replicated plots. The replicated plots were separated by forming ditches of 1.0 m wide. Similarly, each replicated plot was divided in to 7 subplots equal to the number of treatments. Each treatment was separated by bunds of size 30cm x 20cm x 15cm (bottom width x top width x height). Thus the entire plot was divided into 21 subplots indicating 21 replicated treatments. The gross study area, net study area and net treatment plot size were 173.88 m^2, 113.4 m^2 and 5.4 m^2 respectively (Fig. 1). Each treatment was planted with 20 plants in two rows (10 plants in each row).Two lateral lines were laid along the two rows of planting lines in each treatment plot at a spacing of 0.6m (lateral spacing) and drippers were fitted in each line at a spacing of 0.45m (emitter spacing) based on the spacing of the tomato crop as practised by the farmers. Fertilizer was applied at the rate of 125, 80 and 100 kg ha^{-1} of N (urea, 46%), P_2O_5 (single super phosphate,16%) and K_2O (Murate of potash, 60%) in splits (25% N, all P_2O_5 and one-third K_2O at basal; 50% N and one-third K_2O at weeding, hoeing and earthling operation; and 25% N and rest one-third K_2O before flowering). Thus the experimental design was randomized block design (RBD) with three replications and seven treatments. Irrigation was given separately to each treatment by providing regulating valve at each junction point of the lateral and sub-main. Similarly the discharge of water in to sub-main was regulated by provision of pressure regulated valves. Pressure of 1 atm. was maintained at the inlet point of the laterals fitted with pressure gauge. Water was supplied to the pipe line systems by 1.5 H.P. submersible pump.

Table 1; Treatment details of the experiment

Treatments	Capacity of emitter	Irrigation technique	Continuous and pulse timings			
			Continuous irrigation time (min)	Pulse irrigation		
				Supply time (min)	Rest time (min)	Total time required (min)
T_1	2 lph	Continuous irrigation	30min	-	-	-
T_2	4 lph	Continuous irrigation	15min	-	-	-
T_3	4 lph	Pulse irrigation	-	15 (7.5 - 7.5)	15	30
T_4	8 lph	Continuous irrigation	7.5min	-	-	-
T_5	8 lph	Pulse irrigation	-	7.5 (3.75 - 3.75)	22.5	30
T_6	16 lph	Continuous irrigation	3.75	-	-	-
T_7	16 lph	Pulse irrigation	-	3.75 (1.875 - 1.875)	26.25	30

Table 2: Soil moisture content (dry wt. basis) due to vertical and lateral movement of water under different treatments

Treatment	Soil moisture content (dry wt. basis), %													
	Distance from emitter point, cm													
	Horizontal distance(cm) / Vertical distance(cm)	-13	-11	-9	-7	-5	-3	0 (Emission point)	3	5	7	9	11	13
T_1	10		0	2.90	5.42	7.60	9.57	9.99	9.62	9.28	6.40	1.50	0.00	
	20			0	4.12	8.12	8.26	8.44	8.24	8.12	4.10	0		
T_2	10		0	3.26	6.97	10.52	10.62	10.80	10.64	10.52	6.97	1.40	0	
	20		0	1.90	6.11	8.47	8.79	9.27	8.97	8.47	4.52	1.30	0	
T_3	10		0	3.80	7.40	10.30	10.65	11.22	10.60	10.20	7.40	3.40	0	
	20		0	1.80	5.80	9.40	9.54	9.80	9.58	9.10	6.50	2.10	0	
T_4	10		0	4.80	9.20	10.40	10.80	11.02	10.90	10.60	9.10	4.50	0	
	20		0	2.20	6.70	9.40	9.76	10.30	9.50	9.20	6.80	2.80	0	
T_5	10	0	1.2	6.40	10.0	10.80	11.20	11.80	11.40	11.10	10.20	6.00	1.40	0
	20		0	4.80	8.40	9.80	10.20	10.80	10.10	9.80	7.60	3.80	0	
T_6	10		0	5.40	8.20	10.40	10.92	11.40	11.05	10.80	8.40	5.30	0	
	20		0	3.10	7.40	9.20	9.58	10.60	9.80	9.40	7.20	3.30	0	
T_7	10		0	6.10	9.20	10.20	10.76	11.60	11.10	10.90	8.20	4.20	0	
	20			0	6.20	8.60	9.80	10.60	9.80	8.60	6.40	0		

$CD_{0.05} = 0.1484$

$SEM \pm = 0.0478$

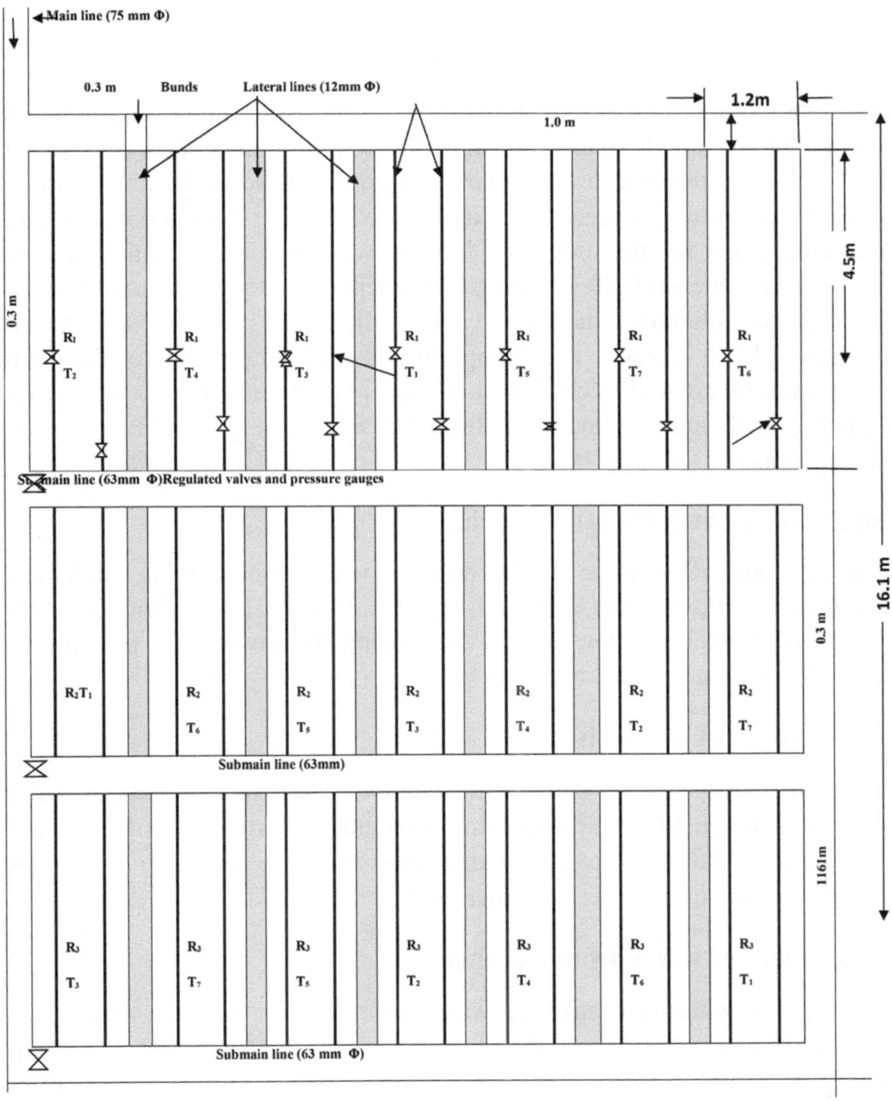

Fig. 1: Schematic diagram of the experimental design

Determinants of the study

Peak water requirement (PWR) of the crop

Under a drip irrigation system, the water requirement includes the crop water demand to meet the losses due to evaporation and transpiration for plant to grow. The peak water requirement (PWR) of the crop was calculated as follows

$$\text{PWR (lit day}^{-1}\text{ plant}^{-1}) = \frac{\begin{pmatrix}\text{Pan evaporation x pan coefficient x crop factor} \\ \text{x canopy cover x plant spacing area}\end{pmatrix}}{\text{Irrigation effieciency}} \quad (3.1)$$

As pan evaporation data of Angul district situated in sub humid region is not available, the data of Bhubaneswar under sub humid climate have been taken for this study. In the month of January, the crop will be in fully vegetative, flowering and fruiting stage. So the average of 10 years maximum day pan evaporation value for the month of January was used in the equation. Pan coefficient was taken as 0.80. The crop factor depend on foliage characteristics, stage of growth, climatic condition etc. Normally 0.9 to 1.1 for closely spaced crops (vegetables) is taken. The canopy implies the shadow area of the plant at 12 noon, which varies according to the growth and remains constant after full plant growth and was taken as 1.0. In drip irrigation system the irrigation efficiency was chosen to be 0.90.

Measurement of emitter flow discharge

Discharge from each emitter in a lateral was measured by placing catch cans under trickle emitters over a period of 10 minutes. The volume of water collected from each emitter in each can during the given time period was measured using a graduated cylinder.

Determination of soil moisture

In trickle irrigation system, soil is wetted in a two dimensional pattern rather than in one dimensional. Soil moisture content in vertical profile (10 and 20 cm) and at different horizontal distances from the emission point was measuredwith the help of digital moisture meter.

Measurement of water front advance

The horizontal waterfront advance was measured with the help of a scale. Vertical waterfront advance was measured by driving the pointed edge of 3 mm diameter and 100 cm long G.I wire at different horizontal distances. Further it was verified by digging the soil around a sample plant from each treatment.

Uniformity coefficient (U_c)

A system with high uniformity coefficient is only possible through properly designed emitters that provide steady discharge to all emission points(Al-Amound, 1995). Uniformity coefficient (U_c), as defined by Christiansen (1942) and modified to reflect a percentage, was calculated using the equation:

$$U_c = \left[1 - \frac{\frac{1}{n}\sum_{i=1}^{n}[q - q_{ave}]}{q_{ave}} \right] * 100 \qquad (3.2)$$

Emission uniformity (E_u)

Merriam and Keller (1978) described the emitter flow variation along the drip pipe line by introducing emission uniformity given as under.

$$E_u = [1 - (q_l/q_{ave})] * 100 \qquad (3.3)$$

Application Efficiency (Ea)

Water application efficiency (Ea) is another important parameter for system selection, design, and irrigation management. The irrigation application efficiency (E_a) was evaluated using following equation described by Zin El-Abedin (2006)

$$Ea = 0.9*E_u - \qquad (3.4)$$

Determination of clogging ratio of emitters

Clogging ratio of emitters is one of the important parameters which used in the evaluation process for pulse drip irrigation performance. The clogging ratio of emitters

"CRE" determined by using the equation of Sultan (1995):

$$CRE = (1 - \eta) * 100 - \qquad (3.5)$$

Where,

Ea = application efficiency (%)

E_u = emission uniformity, per cent

C_v = manufacturer's coefficient of variation

CRE = Clogging ratio of emitters (%)

n = number of emitters per plant for trees and shrubs

η = efficiency of emitter and $\eta = (q_{used} / q_{new})$

q = individual emitter flow rate, lph

q_{ave} = average of discharge rate of all the emitters evaluated in the lateral line, lph

q_l=average of the lowest quarter of the observed discharge values, lph

q_{new} =average discharge for new emitters, lph

q_{used} =average discharge for used emitters, lph

Biometric Characteristics and yield

Biometric characteristics of tomato crop such as plant height, tap root diameter, diameter of stem at base, leaves area per plant, root volume, fruit size etc were observed for the selected plants under different treatments. Diameter of stem at base, tap root diameter and fruit size was measured with the help of screw gauge and slide caliper. Leaf area meter was used to measure the leaves area of tomato plant. Root volume was measured by deepening the entire rooting system of the individual plant in a 1000ml measuring cylinder filled with water. The volume of water displaced will indicate the volume of roots of corresponding plant.Yield of tomato from each treatment were collected from the farmer after each harvest during the growing period.

Determination of water use efficiency

Water use efficiency of tomato yield was calculated according to James (1988) as follows:

$WUE_{tomato,}$ kg m^{-3} = (Total yield, kg / Total irrigation water applied during the crop period, m^3)

Results and Discussion

The results obtained from the field experiment conducted in the study area for the selected crop (Tomato c.v. Chiranjibi) under different physical features and controlling parameters are presented and discussed below.

Computation of peak water requirement

The average of 10 years highest day pan evaporation for the month of January considered for computation of peak water requirement was found to be 3.90 mm. Pan coefficient value was assumed to be 0.8. The crop factor at mid season stage (which is considered as maximum) was taken as 1.1 (Michael, 2009). Based on plant spacing area (0.60 m x 0.45 m) and canopy factor1.0, the peak water requirement was estimated to be 1.03 lit day^{-1} $plant^{-1}$ with irrigation efficiency of 0.90.

Soil moisture distribution

The wetting front and moisture distribution depends on discharge rate and application time. The moisture distribution at different horizontal distances and at 10cm and 20cm vertical depths from trickling point were observed with the help of digital moisture meter are presented in Table3.

Vertical and lateral moisture distribution patterns were found to be different under different treatments. This may be attributed to the different irrigation techniques, flow rates of emitters distance between source and sink and evaporative demands of crops during their successive phonological stages.Soil moisture was decreasing from the emitter point towards the periphery of the wetting front horizontally and also vertically downward. Near and below the dripper point, presence of moisture was maximum and decreased vertically downward. Pulse irrigation resulted in higher values of soil moisture trend for either at 10 cm or at 20cmsoil depths than that of continuous irrigation. When comparison was made between continuous and pulse system of trickle irrigation with the same flow rate, pulsing water discharge produced favourable soil moisture regime by maintaining higher soil water content than continuous flow. Pulse discharges equivalent to continuous discharges increased the soil moisture below the emission point by 3.88% and 5.7% in 4lph; 7.07% and8.7% in 8lph; and 1.6% and 1.92% in 16lph emitters at 10 and 20 cm soil layer depth respectively. In pulse system, amount of water supplied in the first pulse is adjusted in the soil underneath during the off period laterally and vertically which helped in observation of higher moisture in the soil after the supply of water in the second pulse. From the figs. 4.2 and 4.3, it is further observed that, T_5 i.e., pulse application of water through 8 lph emitters resulted better distribution of moisture in the soil as compared to other treatments and was followed by T_4 and T_7 respectively.

Wetting front advance

The lateral and vertical movement of water was varying in all the treatments because of the techniques applied in drip irrigation (Table 3 and Fig. 2). The lateral and vertical movement increased with increase in discharge rate. The lateral movement in pulse system is less as compared to continuous irrigation system irrespective of capacities of the drippers supplying same quantity of water and so also the reverse in vertical downward direction. Continuous application of water makes the tendency of capillary predominance effect at small discharge against the gravity which is more important with higher discharge. The same result has been obtained by Gupta (1995). However, amongst the treatments, T_5 was giving better environment with better distribution of moisture and air in the soil which helped better root growth in the soil and conversely the yield. In treatment T_6, the dripper used is of 16 lphcapacity and water is supplied continuously for a short time period. This made the water spread laterally more in comparison to other application methods.

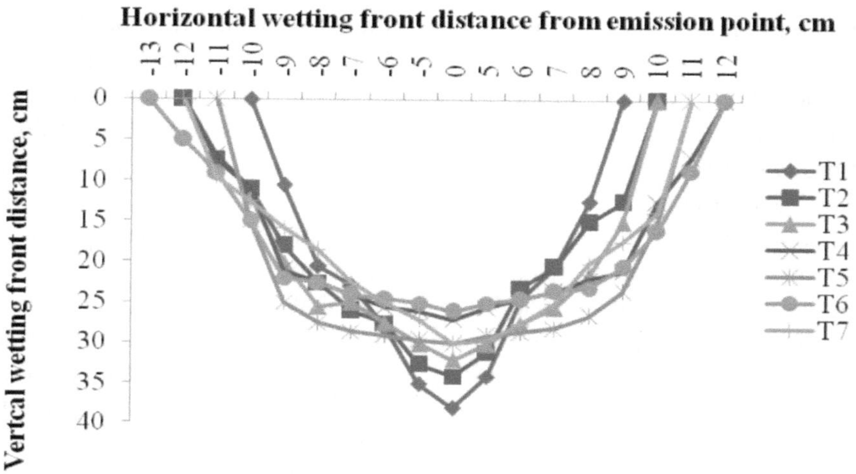

Fig. 2: Wetting front advance under various treatments

Percolation losses (Dp)

In this study, vertical movement under different treatments were recorded and has been presented in Table 3which indicates that the wetting of soil in vertically downward is slightly high in case of T_1 (38 cm)in comparison to other treatments. Tomato crops having roots spreading in the soil within 26 to 38 cm depth, hence deep percolation losses would be practically negligible for such crop as the effective root zone depth for such crop is 60 cm. However deep percolation could also be controlled through proper management and application rates of emitters.

Clogging ratio of emitters (CRE)

Clogging ratio of emitters (CRE)under different treatments was calculated following eqn. 3.5 through the observed data on average discharge of used emitters at initial phase and after completion of harvest period and the results obtained are given in Table 4 and is also shown in Fig. 3.

Table 3: Lateral and vertical wetting front advance with different treatments.

Treatment		Soil moisture content (dry wt. basis), %												
		Distance from emitter point, cm												
	Horizontal distance (cm) / Vertical distance (cm)	-13	-11	-9	-7	-5	-3	0 (Emission point)	3	5	7	9	11	13
T_1	10		0	2.9	5.42	7.60	9.57	9.99	9.62	9.28	6.40	1.50	0.00	
	20			0	4.12	8.12	8.26	8.44	8.24	8.12	4.10	0		
T_2	10		0	3.26	6.97	10.52	10.62	10.80	10.64	10.52	6.97	1.40	0	
	20		0	1.90	6.11	8.47	8.79	9.27	8.97	8.47	4.52	1.30	0	
T_3	10		0	3.8	7.40	10.30	10.65	11.22	10.60	10.20	7.40	3.40	0	
	20		0	1.8	5.80	9.40	9.54	9.80	9.58	9.10	6.50	2.10	0	
T_4	10		0	4.8	9.20	10.40	10.80	11.02	10.90	10.60	9.10	4.50	0	
	20		0	2.2	6.70	9.40	9.76	10.30	9.50	9.20	6.80	2.80	0	
T_5	10	0	1.2	6.4	10.0	10.80	11.20	11.80	11.40	11.10	10.20	6.00	1.40	0
	20		0	4.8	8.40	9.80	10.20	10.80	10.10	9.80	7.60	3.80	0	
T_6	10		0	5.4	8.20	10.40	10.92	11.40	11.05	10.80	8.40	5.30	0	
	20		0	3.1	7.40	9.20	9.58	10.60	9.80	9.40	7.20	3.30	0	
T_7	10		0	6.1	9.20	10.20	10.76	11.60	11.10	10.90	8.20	4.20	0	
	20			0	6.20	8.60	9.80	10.60	9.80	8.60	6.40	0		

Table 4: Clogging ratio of emitters under different treatments

Treatments	q new	q used	CRE(%)
T1	1.87	1.72	8.02
T2	3.65	3.44	5.75
T¬3	3.66	3.46	5.46
T4	7.84	7.37	5.99
T5	7.85	7.40	5.73
T6	15.84	15.48	2.27
T7	15.85	15.54	1.96

Table 5: Uniformity Coefficient (Uc) and Emission uniformity (Eu) and Application efficiency(Ea) for various treatments.

Treatments	Uniformity coefficient, Uc (%)	Emission uniformity, Eu (%)	Application Efficiency, Ea (%)
T1	97.00	91.20	82.08
T2	96.90	91.90	82.71
T3	97.40	92.30	83.07
T4	98.20	95.70	86.13
T5	98.50	96.80	87.12
T6	97.30	92.90	83.61
T7	97.60	93.50	84.15

Table 6: Biometric characteristics of tomato crop under different treatments

Treatm-ments	Stem diameter (cm)	No.of leaves plant-1(nos.)	Leaf area (cm2)	Root volume (cm3)	Max. root diameter (cm)	Fruit diameter (cm)	Fruit wt. (gm)	Ht. of tree (cm)
T1	1.85	180	34.58	27.43	0.50	4.27	63.13	75.00
T2	2.04	192	35.14	30.80	0.60	4.72	63.49	76.87
T3	2.04	194	46.57	30.65	0.70	4.85	65.02	77.80
T4	2.13	197	70.96	34.96	0.98	5.30	66.99	84.67
T5	2.22	200	75.94	35.40	1.06	5.45	67.18	86.07
T6	1.85	190	55.87	33.00	0.78	5.02	65.47	80.17
T7	2.07	194	56.06	34.40	0.90	5.23	66.27	81.17
CD 0.05	0.12	3.56	1.20	1.14	0.12	0.18	0.53	1.04
SEM±	0.03	1.15	0.35	0.33	0.04	0.05	0.15	0.30

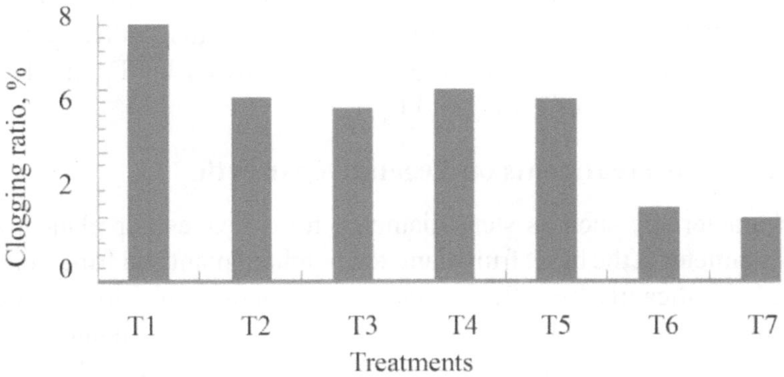

Fig. 3: Effect of different tretments in surface drip irrigationon clogging ratio of emitters

CRE decreased with increasing the capacity of emitters. Less was found in T_6 and T_7. Less value of CRE was observed in pulse application of water in comparison to continuous supply of water through the same emitter supplied with same equivalent quantity of water. Pulse technique (cycle on-off) of water application created turbulence in the flow channel and this turbulence prevent particles suspended from accumulation in flow channel and outlets of emitters. On analogous, this result is in agreement with the results of Al-Amoud and Saeed (1988) and Jackson and Kay (1987, Abdelraouf *et al.*,2012). In the case of T_6 and T_7, the creation of turbulence with high flow rate and flow of water through the wider flow path inside the emitter prevent the accumulation of suspended particles though application time is for a small time period. That is why the low CRE values were found when water was applied in continuous and pulse technique through 16 lph emitters.

Uniformity Coefficient (U_c), Emission Uniformity (Eu) and Application Efficiency (E_a)

Uniformity coefficient (U_c) and emission uniformity (Eu) and application efficiency(E_a) for different emitters under different treatments were calculated through the eqns.(3.2), (3.3) and (3.4) respectively and the results are given in the Table 5.The highest value of U_c (98.50 % and Eu (96.80 %) were observed in T_5 and the lowest being in T_1 with $U_c$andEu values of 97.00 % and 91.20 % respectively. It is also observed that U_c, E_u, and E_a are increased due to pulse application than continuous supply of water through the same emitter supplied with same equivalent quantity of water. This is due to positive effect of CRE. On analogous, this result is in close agreement with Bader et. al. (2001). The estimate showed that U_c and Eu of all the emitters irrespective of supplying water continuously or in pulse are more than 90.00 %. Also the calculated E_a values are more than 80.00 %. The high $E_a$is observed in T_5 followed by T_4 and

T_7 respectively. Amongst all the treatments, pulse application ofwater through 8 lph (T_5) indicated better system of application of water than other systems supplying same quantity of water to tomato crop followed by T_4 and T_7 as indicated by the parameters of U_c, E_u, and E_a.

Effect of Irrigation Treatments on Vegetative Growth

Biometric characteristic such as stem diameter, no of leaves per plant, root volume, root diameter at the base, fruit diameter, height of plant and fruit weight were changed significantly by different application method of drip irrigation under different treatments(Table 6). The maximum value of stem diameter was obtained in T_5 (2.22 cm) which is 20 %, 8.82 %, 8.82%, 4.22% 20.00%, and 7.20% higher than T_1, T_2, T_3, T_4, T_6, and T_7 respectively. Similarly for other parameters maximum value was obtained in T_5 that is pulse application of water through 8lph emitter discharging same equivalent amount of water.

Plant height was significantly affected by treatments. Treament T_5 produced the highest plant growth followed by treatment T_4 and T_7. The lowest value was recorded in treatment T_1.

The leaves areas were significantly affected by treatments. The highest value was observed in treatment receiving high discharge pulse irrigation (T_5) than those followed by T_7 and T_6. The lowest value was recorded in T_2 and T_1.

The root volume 35.4 cm^3 was obtained in T_5 in comparison to other treatments supplied with same equivalent amount of water. Significant differences in fruit fresh weight, fruit diameter were found among treatments (Table 4.4). The best value was observed in T_5 followed by T_4 and T_7. This indicated that in T_5 in which the average fruit weight is more and so also resulted the higher yield of tomato.

Significant differences in fruit fresh weight, fruit diameter were found among treatments (Table 4.4). The best value was observed in T_5 followed by T_4 and T_7. This is indicated in T_5 in which the average fruit weight is more and so also resulted the higher yield of tomato.

Yield and water use efficiency of tomato under different treatment practices

The yield of tomato is presented in Table 4.5 showed significant difference underdifferent treatment practices. Highest yield of 405.33 q/ha was recorded in T_5i.e., pulse irrigation through 8lph emitter. Treatments T_4 and T_7 were the 2nd and 3rd highest yielder in which theyield of tomato was observed to be 392.17 and 386.18 q/ha respectively. The increase of yield due to T_5 was

10.15, 8.77, 7.62, 3.35, 7.39, and 4.95 per cent higher over treatments T_1, T_2, T_3, T_4, T_6, and T_7 respectively. It is also observed from the table that due to better distribution of moisture and air in the soil, pulse irrigation exhibited higher yields when compared with continuous supply of water through the drip emitters irrespective of the capacity supplying same quantity of water to the plants.

High water use efficiency is obtained in T_5 (7.87) followed by T_3 (7.82) and T_2 (7.78). Water use efficiency in the case of T_4 and T_1 is observed to be at par. Less values are obtained in T_6 and T_7. The values were 7.13 and 7.30 respectively. Considering yield, payback period and the farmer's view ignoring the water use efficiency, T_5 followed by T_4 and T_7 are the suitable methods for application of water to tomato crop.

Economics in cultivation of tomato under various treatments

The layout cost of drip irrigation system based on the requirement of different components per hectare cultivation of tomato crop along with the cost economic parameters under different treatments are presented in Table 8. From the table it is found that the NPV value varying depending on the treatment. The NPV values of the treatments were estimated to be 10,05,569, 10,36,565, 10,61,954, 11,67,395, 12,47,792, 11,23,257 and 11,61,372 rupees respectively. Highest being estimated in treatmentT_5 and least in treatment T_1.

Similarly high benefit cost ratio of 3.17 was observed in treatment T_5 and 2.76 in treatment T_1. The payback period (PBP)was found to be 3.39, 3.32, 3.36, 3.07, 2.91, 3.34 and 3.16 years when tomato is irrigated by drip irrigation with treatments T_1, T_2, T_3, T_4, T_5, T_6 and T_7 respectively. The payback period was found less in treatment T_5. The second and third lowest payback period in cultivation of tomato was found in treatments T_4 and T_7 respectively.

Table 7: Yield and Water Use efficiency of Tomato as affected by different water application technique

Treatments	Yield(q ha-1)	Total quantity of water applied (m3)	Water use efficiency (Kg m-3)
T1	367.96	4839	7.6
T2	372.64	4785	7.78
T3	376.63	4812	7.82
T4	392.17	5137	7.63
T5	405.33	5150	7.87
T6	377.43	5292	7.13
T7	386.18	5285	7.3
CD 0.05	0.54		
SEM±	1.87		

Table 8: Computed cost economic parameters under different treatments

Treat-ments	Fixed cost (Rs.)	Variable cost (Rs.)	Present worth variable cost (Rs.)	Total cost out flow (Rs.)	Total discounted scash inflow (Rs.)	NPV(Rs)	Benefit-cost ratio	Payback period
T1	355506	304098	216908	572414	1577984	1005569	2.76	3.39
T2	355506	302417	215693	571199	1607764	1036565	2.80	3.32
T3	355506	302417	215693	571199	1633153	1061954	2.85	3.36
T4	355506	301932	215343	570849	1732038	1167395	3.02	3.07
T5	355506	301932.	215343	570849	1818641	1247792	3.17	2.91
T6	355506	301678	215159	570665	1693922	1123257	2.86	3.34
T7	355506	301678	215159	570665	1732037	1161372	2.96	3.16

Conclusion

From the study, the following conclusions were drawn: Soil moisture decreased from the emission point towards the periphery of the wetting front both horizontally and vertically downward. Near and below the emitter point, presence of moisture was maximum and decreased vertically downward. Pulse irrigation resulted higher soil moisture at 10 and 20 cm soil depths. Lateral movement of water on the soil surface was less than the vertical movement in all types of emitters supplying same quantity of water. The lateral and vertical movement of water increased with increased flow rate of emitters. Lateral movement in pulse application of water is less as compared to continuous supply irrespective of the capacities of the drippers supplying same equivalent amount of water and the reverse in vertical downward direction. The maximum vertical movement of water was 38 cm and effective root zone depth of tomato crop is 60 cm, so percolation loss was not observed in any of the treatments. The maximum vertical movement of water was observed in the treatment T_1 i.e., irrigating the crop though 2lph emitters continuously. The uniformity coefficient was observed more than 90% for all the types of emitters supplying water continuously or in pulse and so also the emission uniformity. The highest being observed in T_5 i.e., water applied in pulse through 8 lph emitters. The E_a is increased up to T_5 and decreased with increasing emitter capacity for both types of application methods (continuous and pulse). Clogging ratio of emitters decreased if the flow rate of emitter is increased up to 8 lph. Further increase in flow rate, CRE is increased for the same equivalent amount of water supply. Better growth of plants was observed in pulse application of water through 8 lph emitter (T_5) followed by T_4 and T_7. Significant difference in yield of tomato was obtained when water was applied through emitters of different capacities emitting water continuously or in pulse method. Yield of tomato was increased due to increased flow rate of emitters up to 8 lph and decreased with increased emitter capacity supplying same quantity of water. The highest yield of tomato was observed in T_5 followed by T_4 and T_7. Highest benefit cost ratio (3.17) was estimated in T_5 followed by T_4 (3.02) and T_7 (2.96). Less benefit-cost ratio of 2.76 was estimated in T_1 (continous application of water with 2 lph emitters). Less payback period of 2.91 years was estimated in pulse application of water through 8 lph emitters. The next less payback period of 3.07 years and 3.16 years were estimated for continuous application of water through 4 lphemitter(T_4) and pulse application of water through 16 lph emitter (T_7). So keeping all in view, treatment T_5 i.e., pulse application of water to tomato crop through 8 lph emitters is considered as the suitable method from yield and economics point of view.

5

Design and Development of Bullock Drawn Mulch Cum Drip Laying Machine

A.V. Kothiya, A.M. Monpara and B.K. Yaduvanshi

Introduction

Agriculture is the most important sector of Indian Economy. The agriculture sector accounts for 18 per cent of India's gross domestic product (GDP) and provides employment to 50 per cent of the countries workforce. India has emerged as the second-largest producer of fruits and vegetables in the world. Formerly agriculture was more dependent on nature and all the operations were carried out manually, however new technologies have been developed to increase the production and productivity of crops (Rangbhal, 2018).

The word mulch has probably derived from the German word "molsch" means soft to decay, which apparently referred to the gardener's use of straw and leaves as a spread over the ground as mulch (Jack *et al.* 1955). Thus, mulching is the process or practice of covering the soil to create favorable conditions for plant growth, development and efficient crop production. Technically, the term 'mulch' is referred to covering of soil (Anon., 2020). Mulching reduces the deterioration of soil by way of preventing the runoff and soil loss, minimizes the weed infestation and checks the water evaporation. Thus, it facilitates for more retention of soil moisture and helps in the control of temperature fluctuations, improves physical, chemical and biological properties of soil, as it adds nutrients to the soil.(Ray and Biswasi, 2016).In drip irrigation, water is applied to each plant separately in small, frequent, precise quantities through dripper emitters. It has high water use efficiency, higher crop yield, less labour requirement and relatively low operating cost, less weed growth, less insect and pest attacks, shorter growing season and earlier harvest of the crop.(Anon., 2020).Mulch can modify soil temperature (Abdul-Baki *et al.*1992).Mulching helps to conserve soil moisture content, also it modifies the physical environment of soil, which helps inappropriate growth of crops (Chakraborty 2008, 2010). The combination of the raised bed, drip irrigation, plastic mulching and fumigation leads to more yield as compared to traditional cropping practices. Even though this technology is quite old and versatile, its use is still limited due to difficulty of laying mulch, drip and hole punching at the required exact position. This is because of the lack of availability of skilled labour and high labour cost. There are some tractor operated and automatic mulch spreading & drip laying machines available in India. These machines were used to prepare soil bed, laying of drip line, laying of plastic mulch film, covering the mulch film on soil by shovels and making holes in mulch film. (Siddesh and Veerangouda 2017, Tipayale *et al*., 2017, Sutar *et al*.2017 and Magdum *et al*.2017).

But, these are not economical due to high cost and limited area under mulching with a drip system. Bullock drawn machine may be more useful for the small-scale farmers those who are growing high yielding variety crops. Still today, a large number of farmers are laying the drip line and mulching sheet manually. For laying, the drip line and mulching sheet require lots of labours because both operations are done separately one after another which increases labour requirement and cost of operations. These problems may be resolved by developing low-cost implements by using multipurpose tool carrier (Vagadia, 2016) for operations of drip and mulch laying. For the small and marginal farmer, it will suitable because of low-cost technology and availability of bullock. Looking to the above fact to overcome the problem of mulch and drip laying operation and to reduce the cost of operation, the attempt has been made to design and develop a bullock drawn mulch cum drip laying machine.

Materials and Method

Bullock drawn mulch cum drip laying machine was developed in the department of Farm Machinery and Power. The experiments were conducted at Polytechnic in Agricultural Engineering, AAU, Muvaliya Farm, Dahod.

Theoretical Consideration of Design

1. Power requirement of the machine should not exceed the power available from bullock. (1hp)
2. Agronomical requirements of the vegetable crops.
3. Field condition during mulch laying.
4. Labour requirements for mulch and drip laying.
5. Ease of operation and maintenance.
6. Safety in operation and the operator's comfort.
7. The expected level of cost of machine and cost of operation.

Functional Requirement

1. Laying of mulch sheet on the prepared bed.
2. Covering the mulch sheet edges with soil from both sides 10 to 15cm.
3. Laying of drip line below the mulch sheet on the prepared bed.
4. Making hole on mulch sheet as per the agronomical requirement of crop spacing.

Agronomical Requirement

Following agronomical requirements were also considered for the design of the machine: Soil bed width = 50 and 80 cm, soil bed height = 10 - 15 cm and width of plastic mulch sheet = 0.60 - 1.20 m.

Economical Consideration

1. The cost of the mulch sheet cum drip line laying machine should be as low as possible so, that small farmer can afford to purchase the machine.
2. The material for the construction of different components should be easily and locally available.

Design and Construction of Bullock Drawn Mulch cum Drip Laying Machine Multipurpose Tool Carrier (Main Frame)

Mainframe consisted of headpiece, beam and clamps. Holes were drilled on the headpiece to mount the other parts of the machine like pressing cum transport wheel, mulch sheet punching mechanism, etc. These holes were made to adjust the working width of the machine according to crop requirement and available mulch sheet width.

The beam was made up of MS round pipe to get better welding strength. MS round pipe of three different diameters was used to make three different sections of the beam with the aim to reduce its weight. A V-shape beam was made to get better stability of the machine.

Mulch roll is available in the market having 0.60 to 1.20 m widths. It was generally available in 400 m mulch length. Thus, the diameter of the mulch roll is varying 150 mm to 250 mm according to the thickness of the mulch sheet. Drip line roll having a diameter of 560 mm and width of 350 mm is commonly available in the market. In a single roll of drip line having a length of drip line is 400 m. The diameter of drip line is 12, 16 and 20 mm.

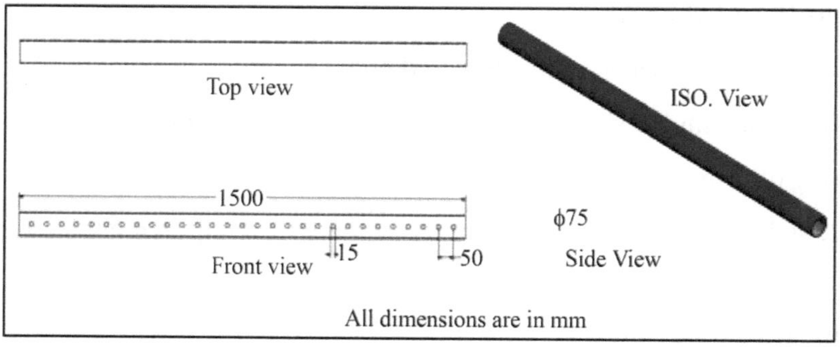

Fig. 1: Detailed drawing of the headpiece

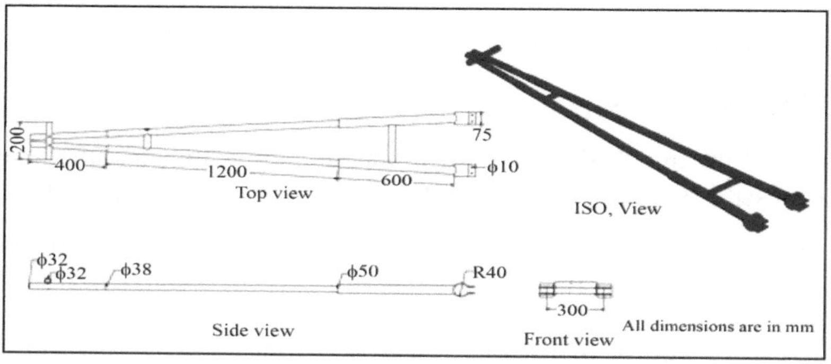

Fig. 2: Detailed drawing of beam

According to the size of the mulch roll and drip line roll, the other components of the machine were designed.

Thickness of central pipe

The thickness of the pipe was calculated by the formula, (Khurmi and Gupta, 2009) M/I = F/Y

Where,

M = Bending moment act on the pipe section, kg-cm

I = Moment of inertia of cross-section of pipe about the neutral axis, cm^4 F = Bending stress, kg/cm^2

Y = Distance from the neutral axis to the extreme edge (cm)

Length of circular pipe would be 15 cm and an outer diameter of the circular pipe would be 32 mm. It would be considered that uniformly distributed load act over the entire length of the pipe. The maximum weight among the selected mulch rolls was 19.5 kg. So, the maximum load on one side would be 9.25 kg. or say it to be 10 kg.

Bending moment, $M = wl^2/2$

$= 10 \times (15)^2 / 2$

$= 1125$ kg-cm

F = Allowable bending stress of mild steel = 800kg/cm^2, (Pandya and Shah, 1986) Y = 3.2/2= 1.6cm

And, $I = \pi / 64 (d^4 - d^4)$ for circular section, (Khurmi, 2006)

$= \pi/64 ((3.2)^4 - d^4)$

Where,

d_o = outer diameter of pipe d_i = inner diameter of pipe Now,M/I = F/Y

$1125/ (\pi/64 ((3.2)^4 - d^4)) = 800/1.6$

d_i= 2.77 cm

Therefore, thickness of pipe = $(d_o - d_i) / 2$

$= (3.2 - 2.77) / 2$

= 0.215 = 2.15 mm

So, it was taken as 2.6 mm as available size of MS pipe.

Thickness of MS flat

Length of the flat would be 15 cm and breadth 2.5 cm. considering all load on the extreme end of the flat.

Bending moment, M = w x l

= 10 x 15

= 150 kg-cm

$I = bt^3 / 12$ for rectangular section, (Khurmi, 2006) Where,

t = thickness of flat, cm

b = width of flat, cm (2.5 cm) and, Y = 2.5/2 = 1.25 cm Now, M/I = F/Y

$150/ (2.5 t^3 / 12) = 800/1.25$

t = 1.04 cm = 10.40 mm So it was taken as 10 mm.

The motion between mulch roll and the carrying unit was sliding. This L shape structure was welded with around clamp. It was mounted on the headpiece and moved according to the width of mulchroll.

This drip line roll carrying unit made according to the size of drip line roll available in the market.

Thickness of drip carrier pipe

Length of circular pipe would be 40 cm and an outer diameter of the circular pipe would be 32 mm. It would be considered that uniformly distributed load act over the entire length of the pipe. The maximum load would be 20 kg.

Bending moment, $M = wl^2/8$

$= 20 \times (40)^2 / 8$

$= 4000$ kg-cm

F = Allowable bending stress of mild steel = 800kg/cm^2, (Pandya and Shah, 1986) $Y = 3.2 / 2$

$= 1.6$ cm

And, $I = \pi / 64 \, (d_o^4 - d_i^4)$ for circular section, (Khurmi, 2006)

$= \pi/64 \, ((3.2)^4 - d^4)$

Where,

d_o = outer diameter of pipe d_i = inner diameter of pipe Now, **M/I = F/Y**

$4000 / (\pi/64 \, ((3.2)^4 - d^4)) = 800 / 1.6$

$d_i = 2.76$ cm

Therefore, thickness of pipe $= (d_o - d_i) / 2$

$= (3.2 – 2.76) / 2$

$= 0.22$ cm $= 2.2$ mm

So, it was taken as 2.6 mm as available size of MS pipe.

Thickness of drip carrier column

The thickness of the pipe was calculated by the formula, (Bansal, 2012) $P = \pi^2 E I / 4 L^2$

Where,

P = crippling load, kg

E = modulus of elasticity of column material, kg/cm^2

I = Moment of inertia of cross-section of pipe about the neutral axis, cm^4 L = length of column, cm

Considering eccentric load on the circular pipe as 10 kg for each column with:

P =10 kg, E = 210 x 10^4 kg/$cm^2$for mild steel, $I = \pi/64\ (d^4 - d^4)$ for circular cross section, and L= 45 cm

$10 = \{\pi^2 \times 210 \times 10^4 \times \pi/64\ ((2.5)^4 - d^4)\} / 4\ (45)^2$

$d_i = 2.48$ cm

Therefore, thickness of pipe = $(d_o - d_i) / 2$

= (2.5 – 2.48) / 2

= 0.01cm = 0.1 mm

So, it was taken as 2.6 mm as available size of MS pipe.

This structure mounted on headpiece using nut and bolts.

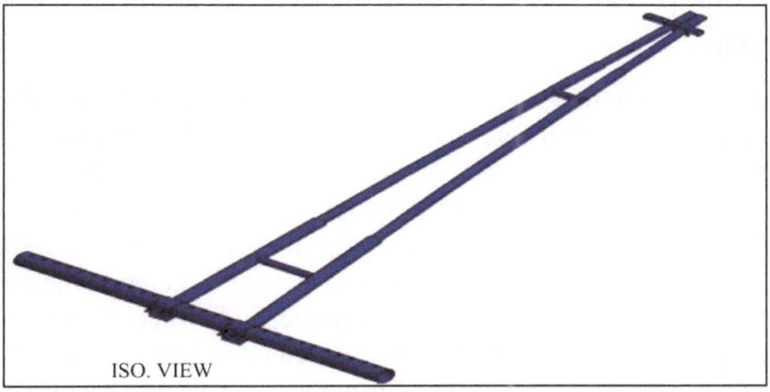

Fig. 3: Isometric view of multipurpose tool carrier (mainframe) Mulch Roll Carrying Unit

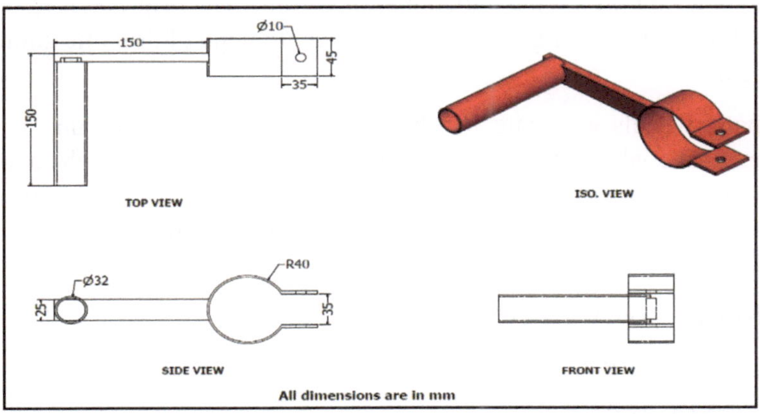

Fig. 4: Detailed drawing of mulch roll carrying unit Dripline Roll Carrying Unit

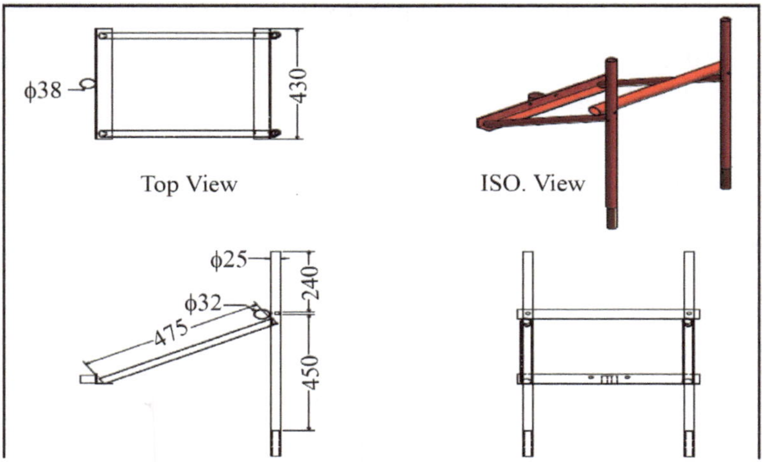

Fig. 5: Detailed drawing of dripline roll carrying unit

Drip Line Guide

A drip line guide made up of wooden material. The diameter of the wooden wheel was kept 150 mm, which was sufficient to guide the drip line without excessive bending. The thickness of this wheel was 40 mm. A rip of the diameter of 25 mm was excavated on the periphery of the guide wheel to accommodate the drip line during operation. Drip line guide wheel mounted on angle iron frame of 30 × 30 ×4 mm. five holes were made on the frame at a distance of 25 mm to adjust the height of the drip line guide wheel. Length of drip line guide frame was 500 mm. Holes were drilled up to half-length of the frame to carry up the guide wheel during transportation. This drip line guide was attached on the big end of the beam by angle iron support . Initial guide was provided on a support having a diameter of 25 mm. The drip line is to be initially passed through it during operation. Detailed drawing of drip line guide is shown in fig.6.

Small Pneumatic wheels were attached behind head piece. Wheels run over the mulch sheet. The wheel starts rotating and at the same time, mulch sheet roll started rotating and unwound. As bullock moves forward, the sheet was also continued unwound and lay along with drip line. It was also working as a transport wheel during on-off field movement of the machine.

$I = 1/3\ (ty^3+b\ (a-y)^3 – (b-t)\ (a-y-t)^3$

$= 1/3\ [0.4 \times 2^2 + 4\ (4 - 2)^3- (4 - 0.4)\ (4 – 2 - 0.4)^3]$

$= 16.12$ kg. cm^2 $M/I = F/Y$

M = Bending moment act on the pipe section, kg-cm

I = Moment of inertia of cross section of pipe about the neutral axis, cm^4 F = Bending stress, kg/cm^2

Y = Distance from neutral axis to the extreme edge, cm 450/16.12 F/2

F = 55.83 kg/cm^2 which is less than the maximum allowable bending stress for mild steel (800 kg/cm^2)

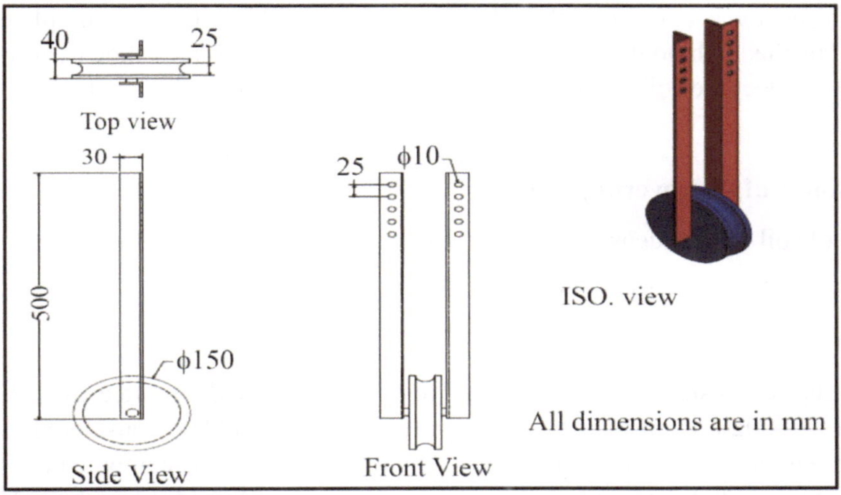

Fig. 6: Detailed drawing of drip line guide Mulch Sheet Pressing cum Transport Wheel

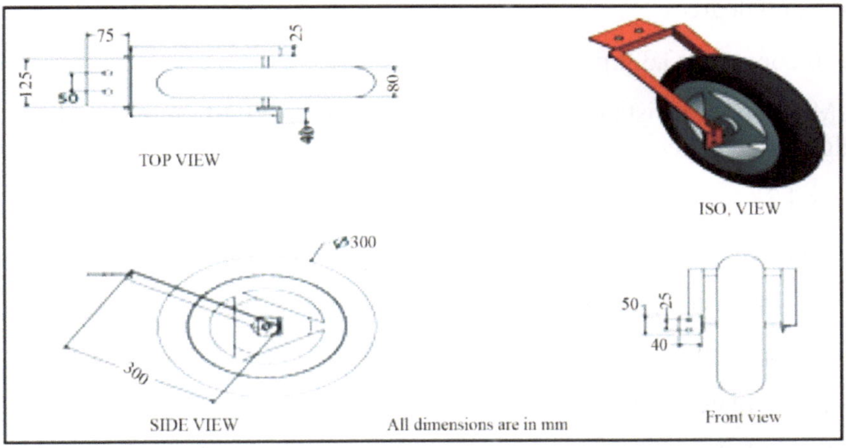

Fig. 7: Detailed drawing of mulch sheet pressing cum transport wheel Mulch

Sheet Covering Plates Stress on wheel frame

Pneumatic wheels having a size (2.75–10, 4PR) were used in the machine for pressing the mulch sheet edges on the ground. The pneumatic wheel was framed in angle iron of 40 × 40 × 4 mm. Length of the frame was 300 mm. A drilled MS flat was welded to the frame for assembling it with the headpiece by nut & bolts. Center to center distance of both wheels was adjustable by changing its position on holes of the headpiece, from 450 mm to 1200 mm. So, different mulch width can be laid by this machine.

Mulch sheet was covered with soil by two covering plates. Conversing plates were attached behind the press wheels at outer sides. The plates slide over the field to collect the soil and push it over the mulch sheet from both sides.

Thickness of the covering plates

Mass of soil = Bulk density of soil x Volume of soil to be conveyed

= 1.27 x (35 x 10 x 15) / 1000

= 6.66 kg

It would be considered that the above mass was evenly distributed over the entire working area of the plate. The volume of soil would be considered for 15 cm length. This mass would act horizontally on plates. The component of it that would act perpendicular to the surface of the plate would cause the B.M. in the plate. The angle between the plate surface and the direction of travel would be 60 degrees.

Force=6.66cos 60=3.33kg

As B.M. = Force ×Distance from C.G of the effective portion of the plate to the edge of the plate

= 3.33 x 4.37

= 14.55 kg-cm

I = bt^3 / 12 for rectangular section, (Khurmi, 2006) Where,

t = thickness of plate, cm

b = width of plate, cm (35 cm)

F = Allowable bending stress of mild steel = 800kg/cm^2, (Pandya and Shah, 1986) Y = 8.75 / 2

= 4.37 cm

Now, **M/I = F/Y**

$14.55 / (35 t^3 / 12) = 800/4.37$

t = 0.3 cm = 3 mm

Therefore, it was taken as 3 mm as the available size of MS sheet. To avoid bending of the sheet, the frame of MS flat is provided for the covering plate which has a thickness of 3 mm. Reinforcement of 25 ×4 mm MS flat if provided behind the active face. Also, the lower edge of the plate was bent forward to avoid bending and ease of gathering the soil.

Covering plates were made of MS sheet having thickness 3 mm. Length and height of plates were 350 mm and 175 mm. Covering plates were attached beside the press wheels at an angle of 60° with the direction of motion. Covering plate was bent 25° at 50 mm above from the bottom edge. These plates were attached at the outer side of both press wheels. Height of the covering plates was kept adjustable. A 40 × 40 × 4 mm angle iron having 245 mm length was used to adjust the height of the plates. Holes were drilled on angle bar at 25 mm spacing for adjustment of height.

Hole making mechanism consisted of the circular wheel, pegs and frame. The circular wheel was made of MS flats having dimensions of 40 × 4 mm. Holes were drilled on the periphery of the wheel to attach pegs. Distance between two holes was kept 100, 150, mm to get different spacing according to the agronomical requirement of crops.

Consider the minimum distance between two holes on the mulch sheet = 30cm on the circular wheel, Number of pegs for minimum distance = 120 / 30 = 4

$120 = \pi \times D$ (D = Dia. of circular wheel) D = 38.21

The form above design it can get 30, 40, 60, 120 cm hole spacing with removable spikes. These pegs were made of MS circular pipe. One end of the peg was sharpened enough to cut the mulch sheet. Another end was welded with MS flat and made a hole at the center of peg to mount it on the circular wheel with help of nut-bolts.

The frame was made of iron angle having a size of 40 × 40 × 4 mm. The frame was joined with movable MS flat for up and down movement of wheel during transportation of machine and field operation. This 120 × 75 mm MS flat with a thickness of 6 mm was mounted on the mainframe using nut-bolts. At the end of the frame, two bolts were attached to put the extra weight that required punching for different thickness of mulch sheet.

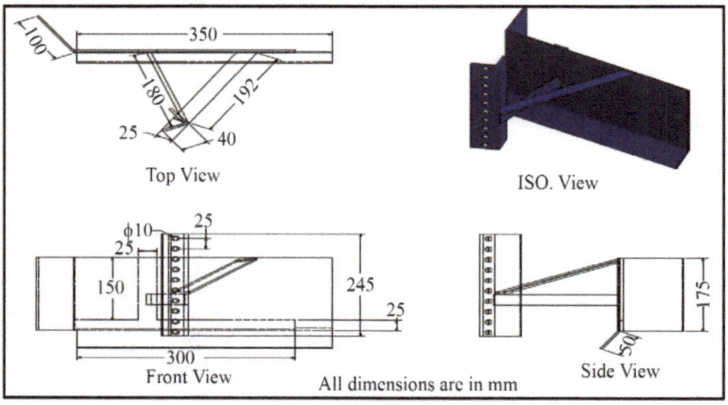

Fig. 8: Detailed drawing of mulch sheet covering plates Mulch Sheet Punching Mechanism

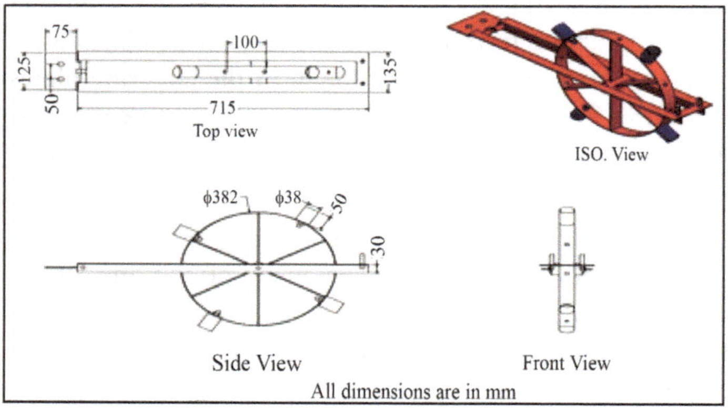

Fig. 9: Detailed drawing of punching making mechanism

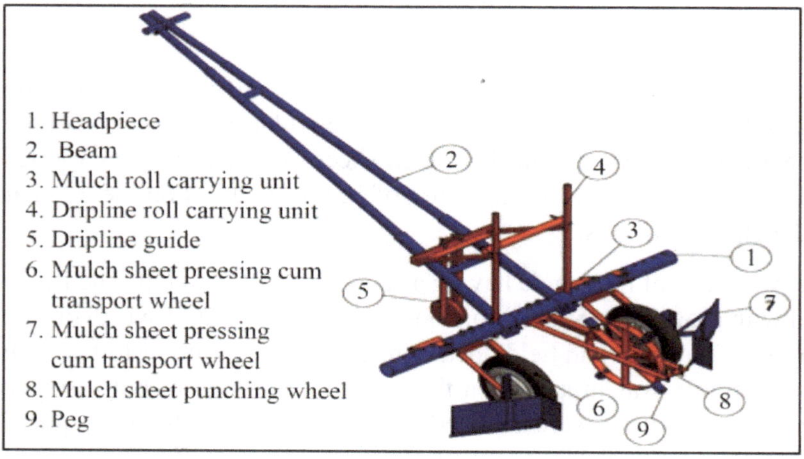

Fig. 10: Isometric view of bullock drawn mulch cum drip laying machine

Fig. 11: Developed bullock drawn mulch cum drip laying machine

Fabrication of machine

Multipurpose tool carrier was the mainframe of the machine. It consisted of a beam, headpiece and clamps. Beam was joined with headpiece with the help of clamps. Transport cum mulch pressing wheels attached on headpiece with help of nut and bolts. Center to center distance between wheels was adjustable. The soil covering plates attached at the side of the wheel. Height of the covering plates was adjustable. Thus, during off-field condition covering plates was not touched to the ground. Mulch carrying units were mounted on headpiece with the help of clamps. According to the width of the mulch roll, the distance between two mulch-carrying units was adjusted. Structure of drip roll carrying unit was attached vertically on headpiece with the help of nut and bolts. Drip roll carrying shaft was bolted on the vertical members of the unit with help of nut and bolts. Thus, drip roll can be easily changed.

Results and Discussion

Design of the machine was made in computer-aided software, Autodesk Inventor. Mulch cum drip laying machine was developed as per the functional requirement. When the developed machine is operated on the field, one end of the drip line is initially fixed at ground. When the machine is moved in the forward direction, the drip line is started to unwind and laid on the soil bed. Drip line is initially passed through the drip line guide wheel and laid below the mulch sheet. Drip roll can be used up to a length of 400 m. Simultaneously, as the machine moves forward, the mulch sheet is pressed by press wheel and started to lay on the soil beds over the drip line. Mulch sheet is covered by the soil through soil covering plates. Both sides of the mulch sheet are covered by soil up to a width of 15 cm. Height of the soil covering plates can be adjusted according to soil requirement to cover the much sheet. Width of operation can be adjusted according to the width of the mulch sheet, which may vary from 0.6 m to1.2 m. Holes are made by punching mechanism at required spacing on mulch sheet. The spacing of holes can be changed according to crop requirement by changing the position of pegs on the periphery of the punching wheel.

Conclusions

The developed machine can be used to lay mulch sheet and drip line simultaneously and holes can be made on mulch sheet at variable spacing. The overall dimension of the developed machine was 3210 × 1500 × 950 mm. Weight of the developed machine is 60 kg. Width of operation can be adjusted for different width of mulch sheet varying between 0.6 m to 1.2 m. Height of soil covering plates and drip guide can be also adjusted according to the height of soil beds.

References

Abdul-Baki, A. Spence, C. and Hoover, R. (1992). Black Polyethylene Mulch Doubled Yield of Fresh- Market Field Tomatoes. Hort. Sci. 27:787–789.

Anonymous,(2020). Dripirrigation. http://agritech.tnau.ac.in/banking/PDF/Minor % 2 0 Irrigation Drip % 20 Irrigation.pdf

Anonymous,(2020). Plastic Mulching for Crop Production. http://agritech.tnau.ac.in/agricultural engineering/ plastic_mulching.pdf

Bansal, R. K. (2012). A textbook of strength of material (5th ed.). New Delhi: Laxmi Publications (P) Ltd.

Chakraborty, D. Garg, R. N., Tomar R. K. (2010). Synthetic and Organic Mulching and Nitrogen Effect on Winter Wheat (Triticum aestivum L) in Semi-Arid Environment. Agricultural Water Management. 97: 738-748

Jacks, C. V., Brind, W. D. and Smith, R. (1955). Mulching Technology Comm.Common Wealth. Bulletin of Soil Science, 49:118.

Khurmi, R. S. (2006). A Textbook of Engineering Mechanics (20th revised ed.). New Delhi: S. Chand and Company Ltd.

Khurmi, R. S., and Gupta, J. K. (2009). Theory of Machine (14th ed). New Delhi: Eurasia Publishing House (PVT.) Ltd.

Magdum, S. S., Patil, S. B. and Mulla, L.B. (2017). Automatic Mulch Paper Laying Machine for Agricultural Application. Journal of Advances in Science and Technology, 13(1):241-244.

Pandya, N. C. and Shah, C. S. (1986). Elements of Machine Design. Charotar Publishing House, Anand.

Rangbhal, A.V. (2018). Performance of Mechanical Plastic Mulch Laying Practices: A Review. International Journal of Engineering Science Invention, 7(3): 42-46.

Ray, M. and Biswasi, S. (2016). Impact of Mulching on Crop Production: A Review. Trends in Biosciences, 9(14): 757-767.

Siddesh, M., and Veerangouda, M. (2017). Development and Evaluation of Tractor Operated Plastic Mulch Laying Equipment. International Journal of Agricultural Engineering, 10(2): 374-378.

Sutar, S., Yamgar, R., Patil, P., Bhosale, S. and Siddha, U. Y. (2017). Advanced Mulching Paper and Drip Laying Machine. Imperial Journal of Interdisciplinary Research, 3(10): 534-536.http://www.onlinejournal.in

Tipayale, A., Salunnke, M. S., Thete, S. U., Thete, T. S. and Thete, S. B. (2017). Advance in Mulching Paper-Laying Machine. International Journal for Scientific Research & Development,5(3):217-219.

Vagadia,V. R., Mohnot, P., Vavliya, P. N. and Yadav, R. (2016. Development and Popularization of Multipurpose Pipe Frame Implement for Adivasi Farmers Of Gujarat. An International e-Journal, 5(1): 21-30.

6

Effect of Methods of Land Preparation on Growth and Yield of Transplanted Rice in North Western Plateau Agroclimatic Zone of Orissa - An on Farm Study

S.C. Senapati, J.K. Pati, A.P. Sahu, J.M.L. Gulati and S.K. Behera

Introdoction

Tillage is a prime activity carried out prior to sowing of a crop with an objective to bring the soil to a favorable tilth to provide a congenial environment for crop growth and yield. The operation of ploughing once and left out for a few weeks followed by a broadcasting of rice seeds by hand and then ploughing and raking together at the same time once gave highly significant rice grain yield (Srisa-ard, 2008). Adoption of the "Sawah" system (bunded, puddle and leveled) in inland valleys in Ghana lead to significant improvement in soil and water management; enhancement in rice grain yield and made sustainable production. Even the practice of Sawah system of cultivation increased local rice production by over 300% because of the mechanisms with intrinsic resistance to erosion (better water control and nutrient management (Buri *et al.*, 2012).

Transplanting of rice is mostly practised in medium and low lands during wet season and during dry season in irrigation commands under puddled condition as it provides greater support for growth and development of rice crop (Mambani *et.al.,* 1990). Puddling performs an important activity in transplanted rice that not only provides a favorable soil condition for ease of transplanting and control of weeds (Cherati *et al.*, 2012) but also decreases the loss of water and nutrients through seepage and deep percolation. The reduced conditions so developed also enhance the nutrient availability (Ghildayal, 1978; Sharma and De Datta, 1985). The study conducted by Nwitel *et al.*, (2016) say that all the Sawah (an Indo-Malaysian word for' padi' refers to leveled rice field surrounded by bunds with inlets and outlets for irrigation and drainage water management system) adopted tillage environments (Bunding-with or without pudling - leveling / little leveling) positively improved both the soil parameters (pH, OC, TN and all the exchangeable bases within the periods of study) and rice grain yield was relatively higher than other tillage practices.

Despite these beneficial effects, necessity of puddling in rice is questioned. Elias, as early as 1969, reported that puddling is one of the most tedious and expensive agricultural practices that also destroys soil aggregates and alters other soil physical properties thus reduces root growth. However, puddling has not been found favourable for rice growth and yield on vertisols and poorly drained soils. The study of Kumar *et al*., (2019) in Indo-Gangetic Plains of India, revealed that maximum yields, net returns and mean infiltration rate (0.10 cm hr–1) and least bulk density in 15-20 cm soil depth were observed in rice-wheat system under the practice of zero-till direct-seeded rice with residue followed by zero-till wheat with residue (ZTDSR +R–ZTW +R). Results from this study also revealed that conventionally tilled (CT) and transplanting of rice could be successfully replaced by adoption of the profitable double ZT –RW system. In the Central Terai region of Nepal, zero-tillage combined with residue retention lowered soil bulk density and pH, enhanced P and K availability to plants and improved productivity of the rice-wheat system (Sah *et al*., 2014).

While puddling has been reported to increase the yield of rice by many authors (VandeGor, 1950; Sakanone and Mizunuma, 1962; Sanchez, 1973; De Datta and Kerim, 1974), others have opined that puddling may not be necessary as it has little effect on rice yields (Melabayard and Buencosa, 1967; Scheltena, 1974; Utomo*et al*., 1985; and Sharma, 1985). Fujihara *et al*., (2013) reported that the practice even could not increase the duration of standing water over the field and found that pond water disappeared more quickly in puddled field than non-puddled field.

Land leveling facilitates uniform distribution of water, and has been reported to save about 50 per cent of irrigation water (Singh *et al.,* 1988). Precision land leveling was also favoured optimum flooding depth that helped in increasing nutrient availability and reduced weed menace (Walker *et al.,* 2003). This also helped in improving rice-wheat cropping system profitable and water saving by 2 % over traditional land leveling in Indo – Gangetic planes (Jat *et al*., 2009). Safran (1975) reported that wet leveling operation increased soil bulk density in the upper 20cm layer and reduced grain yield by 8.12% as compared to control. It was attributed to poor distribution and penetration of rice roots under puddled soil.

Owing to the complexity of soil condition, and limited availability of water, the efficient use of irrigation water attains greater importance. Different types of agricultural tools used in farming for initial land preparation, which, among the crop production factors, contribute up to 20% towards the yield (Khursid *et al.*, 2006). During the process, subsoil compaction in the form of plough pan by tillage system, especially through mechanical practices, normally occurs. In Indo-Gangetic Plains of India, maximum infiltration rate (0.10 cm hr^{-1}) and least bulk density in 15-20 cm soil depth were observed inrice-wheat system under the practice of zero-till direct-seeded rice with residue followed by zero-till wheat with residue (Kumar *et al.*, 2019).Zero-tillage combined with residue retention lowered soil bulk density and pH, enhanced P and K availability to plants and improved productivity of the rice-wheat system (Sah, 2014). Further, migration of fine soil particles along with percolating water can result in formation of clay pan in low land rice cultivation. Such altered soil environment, thus, retards functioning of rice roots which affects the overall growth and yield of crop. Such situation, developed over the years of rice cultivation, needs deep ploughing to break the clay pan to facilitate the downward entry of water.

Keeping this in view a study on the effect of methods of land preparation on growth and yield of transplanted rice was conducted under farmers' field during dry season in Pitamahal Irrigation Project of district Sundargarh, Odisha..

Materials and Methods

Study was carried out in farmer's fields during two consecutive dry seasons of 2016 - 17 and 2017 - 18 in Pitamahal irrigation command, North Western Plateau Agroclimatic Zone of Odisha. The experimental site is located in Lathikata block of Sundargarh district, Odisha at 22^0 05' to 22^0 20' N latitude and 84^0 42' to 84^0 55' E longitude. The soil of the study area was sandy clay loam having sand, silt and clay content of 56.5, 11.5 and 32.0 per cent, respectively. The treatments consisted of ploughing by bullock drawn *desi* plow with leveling of 0.50 – 1.00 % slope (T_1), deep ploughing (20 -25 cm) by bullock drawn M.B. plow followed by leveling 0 - 0.25 % slope (T_2) and farmer's practice of ploughing with *desi* plow with leveling 1.00 – 3.00% slope (T_3) were tried in a randomized block design with seven replications. In all the treatments 21 days old rice (cv. Khandagiri) seedlings were transplanted at a recommended spacing of 15 cm x 15 cm. A uniform dose of 80 kg N, 40 kg P_2O_5, and 40 kg K_2O was applied to each treatment in the form of urea, single superphosphate and muriate of potash, respectively. Urea was applied in three splits i.e., 25 per cent at the time of transplanting, 50 per cent at tillering and the rest 25 per cent at panicle initiation stage while all the phosphorous and potash were applied as basal. All other practices were remained common to all the treatments.

Bulk density of soil at 0 - 15 cm and 15 - 30 cm were determined four times i.e., at 1 day after puddling (initial), 15 & 30 days after puddling and at harvest of the crop by core method. Depth of puddling was measured after 24 hrs by gentle pushing a wooden scale in the mud until it hit the hard ground in all the treatments. Initially 2 cm ponding water was maintained in the field up to the time of first top dressing. Thereafter, recommended practice of application of 7 cm water one day after the disappearance of ponded water was followed. Irrigation was withheld 10 days before the harvest of crop (Anonymous, 2001).

Seepage and percolation rate was estimated by monitoring the depth of water regularly at 08.00 am daily. The depth of water during irrigation and ponding period was observed by installing graduated stick at four corners of each treatment plots. The evapotranspiration (ET) value for the specified period was subtracted from the total depth of water lost from the field under the specific treatment to obtain the seepage and percolation loss. The ET value of paddy crop for the period from transplanting to 10 days before harvest was obtained by multiplying the reference crop evapotranspiration with crop coefficient (Doorenbos *et al.*, 1977). In the absence of the pan evaporation data in the nearby area, the data of Regional Research Technology Transfer Sub – Staion, Kirei, Sundargarh were considered for computation of reference crop evapotranspiration and the total ET value is presented in Table 1.

Table 1: ET of rice crop during growth period (15.01.2017 – 19.04.2017).

Month/Number of days	Evaporation (mm)	Crop coefficient	ET (mm)
January (17 days)	57.90	1.10	63.70
February (28 days)	139.30	1.10	153.2
March (31 days)	154.30	1.25	192.90
April (19 days)	140.14	1.00	140.14
Total ((% days)	491.64		549.94

Root samples were collected from each treatment during both the years following trench method for observations on root growth. The average length and diameter of the roots were measured with the help of measuring scale and screw gauge, respectively. The volume of roots hill^{-1} was determined by water displacement method. Plant height, number of tillers m^{-2} and grains panicle^{-1} were recorded at harvest. The total yield from each treatment was recorded after threshing, winnowing and drying of rice grains at 12 per cent moisture. The additional cost involved due to imposition of treatments was recorded at the time of puddling. The data were statistically analyzed following standard procedures (Panse and Sukhatame, 1985).

Results and Discussion

Bulk density

Initially the bulk density at the surface layer of 0 - 15 cm as well as the sub surface layer of 15 - 30 cm soil depth was the lowest in T_2 followed by T_1 and T_3 (Table 2) indicating various degrees of soil compaction in farmer's field. Decrease in bulk density due to puddling has also been reported by Sharma *et al.,* (1988). At subsequent observations, the bulk density increased and reached its highest values of 1.71, 1.80 and 1.61 in 0 - 15 cm layer and 1.78, 1.82 and 1.66 in 15 - 30 cm soil layer in case of T_1, T_2 and T_3,an increase of 33.59, 63.63, 21.96% and 39.06, 65.45, 25.75% under surface and subsurface layers, respectively. The bulk density at sub surface was higher in all the observations than the surface layer. It might have occurred because of breakdown of the soil aggregates due to puddling and subsequent generation of open structure. In the treatments T_1 and T_3, ploughing by *desi* plow perhaps shattered the bottom layer to little depths during preparatory tillage operation resulting thereby a higher bulk density as compared to surface layer though the puddling depth was within 10 - 15 cm. The higher bulk density at surface layer obtained under T_3 is found to be higher as compared to T_1.It may be attributed to improper leveling and uneven depth of water level over the field which might have encouraged more of lateral movement of water than the vertical one that make the soil more compact at surface layer as the finer particles accumulated in the pore spaces. In T_2, the puddling depth varied from 20 - 25 cm, the sub soil layer was well shattered and pulverized. By the time of harvest, soil got compacted with finer particles settling down and finally adjusted in the pore spaces, and thus a relatively higher value of bulk density was observed

Table 2: Effect of methods of land preparation on bulk density of soil (gm cm^{-3}).

Treatment / Soil depth	1 DAP		15 DAP		30 DAP		At harvest	
	0-15 cm	15-30 cm	0-15 cm	15-30 cm	0-15 cm	15-30 cm	0-15 cm	15-30 cm
T_1	1.28	1.32	1.35	1.38	1.62	1.65	1.71	1.78
T_2	1.10	1.15	1.40	1.48	1.65	1.68	1.80	1.82
T_3	1.32	1.38	1.46	1.52	1.57	1.61	1.61	1.66
C.D. (P=0.05)	0.03	0.01	0.036	0.03	0.027	0.11	0.015	0.02

'DAP'- Days after planting

Seepage and percolation

Puddling by M.B. / *desi* plow showed a significant effect on seepage and percolation of water in the experimental fields. The evapotranspiration (ET) value for the specified period (Table 1) was subtracted from the total depth of water lost from the field under the specific treatment to obtain the seepage and percolation loss. The losses were higher in T_2 as compared to T_1 and T_3. This is possibly due to deep ploughing with M.B. plow which led to loosening of the soil (Table 3). On the other hand in the case of T_1 and T_3, the soil was opened up to a small depth and the sub soil layer was not much disturbed during ploughing. The average values of water loss due to seepage and percolation were estimated as 5.11, 7.44 and 5.71 mm day^{-1} in T_1, T_2 and T_3 respectively (Table 3). Ploughing to a greater depth in T_2, perhaps facilitated internal drainage laterally and vertically (seepage and percolation) as indicated by presence of relatively less reddish tinge on soil surface, which is considered as a favourable sign of decline in iron concentration.

Table 3: Seepage and percolation as affected by different methods of land preparation in transplanted rice field.

Treatment	Water applied (cm)	ET crop (cm)	Seepage and percolation between transplanting to 10 days before harvest (cm)	Mean seepage and percolation (mm day^{-1})
T_1	106.40	54.99	51.41	5.71
T_2	122.00	54.99	67.01	7.44
T_3	101.00	54.99	46.01	5.11
C.D. (P=0.05)			2.20	

Plant growth and yield

Plant height, number of effective tillers m^{-2}, number of grains panicle^{-1}, and 1000 grain weight were recorded and are presented in Table 4. Results revealed that the plant height (72.20 cm), number of effective tillers m^{-2} (423.40) and grains panicle^{-1} (71.80) were significantly superior in T_2 compared to other two treatments. The number of effective tillers m^{-2} and grains panicle^{-1} were found to be the deciding factor for yield of rice.

Table 4: Effect of methods of land preparation on growth of rice.

Treatment	Plant height (cm)	Number of effective tillers m^{-2}	Number of grains panicle^{-1}	1000 grain weight (g)
T_1	68.14	390.30	65.90	14.80
T_2	72.20	423.40	71.80	15.10
T_3	63.57	387.7	63.90	14.40
C.D. (P=0.05)	4.98	37.18	11.70	0.30

The root growth parameters under different practices of puddling showed a significant variation (Table 5). Root length (23.88 cm) and volume(81.75 cm^3) were higher in T_2 compared to T_1 and T_3 as ploughing by M.B. plow helped to break down the sub soil layer that provided lesser resistance for the growth of the roots because of reduced compaction. Further, uniform distribution of water over the soil also helped in uniform root growth throughout the profile. Similarly, when comparison is made between T_1 and T_3, more root length (23.49 cm) and volume (69.30 cm^3)were observed in T_1 which might be due to uniform distribution of water as the land was properly leveled after ploughing with *desi* plow. Land leveling influenced the root length and it was more in T_1 (by 5.10 %) and T_2 (by 6.85 %) as compared to T_3 (22.35 cm). This might be due to favourable soil condition particularly at initial stage of the crop growth.

Table 5: Root growth as affected by different methods of land preparation.

Treatment	Root length (cm)	Root diameter (mm)	Root volume (cm^3)
T_1	23.49	0.35	69.30
T_2	23.88	0.36	81.75
T_3	22.35	0.33	56.15
C.D.(P=0.05)	0.06	0.04	9.67

The grain yield of rice (Table 6) ranged from 35.60 in T_3 to 46.20 q ha^{-1} in T_2 across the seasons. Ploughing with *desi* plow and leveling (T_1) significantly increased the rice yield by8.68 % and deep ploughing with M.B. Plow followed by leveling (T_2) was superior by 28.21% over farmers' practice (T_3). Straw yield in $T_1$and T_2 also followed similar trend with 7.21 and 29.58% yield enhancement over farmers' practice (T_3). The practice of land leveling after ploughing for transplanted rice helped in proper distribution of standing water in the field of T_1 and T_2. Proper land leveling might have helped to increase nutrient availability and weed suppression as observed by Walker *et al.*, (2003). Deep ploughing by breaking the existing hard pan in the sub- surface layer facilitated drainage by seepage and percolation which was reflected in the crop growth and ultimately in the grain and straw yield (Table 4, 5 and 6). Further in T_2, farmers observed a decrease in reddish tinge on the soil surface compared to other treatments. Thus, a decline in iron toxicity in treatment T_2 might have casted favourable influence on crop growth and yield.

Table 6: Effect of methods of land preparation on yield of rice.

Treatment	Yield (q ha^{-1})					
	Grain			Straw		
	2016-17	2017-18	Mean	2016-17	2017-18	Mean
T_1	38.40	39.20	38.80 (8.68)*	43.40	44.30	43.85 (7.21)*
T_2	45.60	46.20	45.90 (28.21)*	52.50	53.50	53.00 (29.58)*
T_3	35.60	35.80	35.70	41.30	40.50	40.90
C.D. (P=0.05)	1.04	1.24		1.95	1.56	

*Values in the parenthesis indicate per cent increase over farmers practice (T_3)

Table 7: Economics as affected by different land preparation methods on transplanted rice cultivation.

Treatment	Mean cost of production (Rs. ha^{-1})	Mean cost of produce (Rs. ha^{-1})			Mean net profit (Rs. ha^{-1})	Benefit : cost ratio
		Grain	Straw	Total		
T_1	43642.00	70422.00	6578.00	77000.00	33358.00	1.76
T_2	45308.00	83308.00	7950.00	91258.00	45950.00	2.01
T_3	42930.00	64800.00	6135.00	70935.00	28005.00	1.65

N.B.: Cost of rice grain and straw – Rs.1815.00 and Rs.150.00 q^{-1} respectively

Table 8: Additional profit as influenced by different land preparation methods in rice cultivation

Treatment	Additional cost of treatment imposition over T_3 (Rs.)	Additional profit gained on treatment imposition over T_3 (Rs.)	Additional return per rupee investment (Rs.)
T_1	712.00	5360.00	7.53
T_2	2378.00	17945.00	7.54

Economics

Deep ploughing and leveling (T_2) showed highest net profit of Rs.45950.00 ha^{-1} in comparison with T_1 and T_3 (Table 7). Leveling alone (T_1) could increase the net profit to a tune of Rs. 5353.00 ha^{-1} over farmers' practice (T_3). In case of treatments T_1 and T_2, the mean additional cost involved ha^{-1} were Rs.712.00 and Rs.2378.00respectively (Table 8). The additional return $rupee^{-1}$ investments were calculated and observed to be at per i.e., Rs.7.53 and Rs.7.54 in respect of T_1 andT_2, respectively. However the mean benefit cost ratio of 2.01 was observed with T_2 followed by 1.76 in T_1 and 1.65 in T_3 (Table 7). In the opinion of farmers, the practice of T_2 although involved extra cost, can be accepted as it can compensate the extra expenditure due to higher yield.

Conclusion

The study revealed that deep ploughing (20 – 25 cm) with bullock drawn M.B. plow followed by proper land leveling was advantageous in terms of yield and was cost effective as well, in red laterite soils having impeded internal drainage with possibility of iron toxicity. Thus, deep ploughing (20 – 25 cm) with M.B. plow followed by leveling in red and laterite soils of North Western Plateau Agroclimatic Zone of Odisha can be recommended.

References

Anonymous. 2001. Research Result of On Farm Adaptive Trial under Agriculture Intensification Programme (World Bank), Regional Research and Technology Transfer Station, Chiplima, Sambalpur, Orissa, India.

Cherati, F. E., Bakshipour, S., Hamid, Z., Elnaz, N.T., and Babatabar, R. 2012. Effects of tillage on weed density and performance of rice in the Mazandaran province of Iran. African J. of Agri. Res., 7 (12): 1915-1918.

De Datta, S.K., and Kerim, M.S.A. 1974. Water and nitrogen economyofrainfed rice as affected by soil puddling. Soil Science Soc. Ame. Proc., 38: 515-518.

Doorenbos, J., and Pruitt, W. O. 1977. Guide lines for predicting crop water requirements. FAO Irrigation and Drainage Paper No. 24, FAO, UNO, Rome, Italy.

Elias, R.S. 1969. Rice production and minimum tillage. Outlook on Agriculture, 6(2): 67 – 70.

Fijihara, Y., Yamada, R., Oda, M., Fuji, H., Ito, O., and Kashiwagi, J. 2013. Effects of puddling on percolation and rice yields in rainfed low land paddy cultivation: Case study in Khammouane province, Central Laos. Agri. Sciences, 4 (8): 360 – 368.

Jat, M.I., Gathala, M.K., Ladha, J.K., Saharawat, Y.S., Jat, A.S., Kumar, V., Sharma, S.K., and Gupta, R. 2009. Evaluation of precission land leveling and double zero-till system in the rice – wheat rotation, water use, productivity, profitability and soil physical properties. Soil & Tillage Research, 105: 112 – 121.

Khursid, K., Iqbal, M., Arif, M.S., and Nawaz A. 2006. Effect of tillage and mulch on soil physical properties and growth of maize. Int. J. Agric. Biol., 8: 593 – 596.

Mambani, B., De Datta, S.K., and Redulla, C.A. 1990. Soil physical behaviour and crop response to tillage in low land rice soils of varying clay content. Plant and Soil, 126: 227-235.

Panse, V. G. and Sukhatame, P. V. 1985. Statistical methods for agricultural workers. ICAR, New Delhi.

Saffan, M. M. 1975. Effect of wet leveling on some physical properties and root system development of rice in clay soil. M.Sc. Thesis, Faculty of Agriculture, Tanta University, Tanta, Egypt.

Sakarone, Y., and Mizunuma, Y., 1962. Effect of harrowing upon the behavior of nitrogen in well drained paddy field with gravel layer. J. Sci. Soil Mannure, 33: 386- 390.

Sanchez, P.A. 1973. Puddling tropical rice soils, II, Effects of water losses, Soil Sci., 115: 303-308.

Sharma, P. K., De Datta, S.K., and Redulla, C. A. 1988. Tillage effect on soil physical properties and wet land rice yield. Agronomy Journal, 80: 34 – 39.

Sharma, P.K., and De Datta, S. K. 1985. Effect of puddling on soil physical and processes, In: Soil Physics and Rice, IRRI, Los Banos, Philippines..

Sharme, P.K. 1985. Physical edaphology research in low land rice –based cropping system. IRRI Saturday Seminar, June 22.

Singh, R. V., Yadav, B.P., Jaiswal, C. S., and Chauhan, H. S. 1988. Effect of land levelling index on irrigation efficiency. J. Agril. Engg., XXV (2): 159-164.

Utomo, W.H., Isalmi, T., and Murdoko, B. 1985. The effect of tillage method on the growth and yield of low land rice. Paper presented at the National Congress of Indonesian Soil Science Society, Bogor, Indonesia.

Walker, T. W., Kingrey, W. L., Street, J. E., Cox, M. S., Oldhaui, J. L., Gerard, P.D., and Xiang Han, F. 2003. Rice yield and soil chemical properties as affected by precison land leveling in alluvial soils. Agronomy Journal, 95 (6): 1483-1488.

Kumar, V., Gathala, M.K., Sahrawat, Y.S., Parihar, C.M., Kumar, R., Kumar Rabin, Jat M.L., Jat, A.S., Mahala, D.M., Kumar, L., Nayak, H.S.,Parihar, M.D., Rai, V., Jewlia, H., and Kuri, B,R, (2019). Impact of tillage and crop establishment methods on crop yields, profitability and soil physical properties in rice–wheat system of Indo-Gangetic Plains of India. Soil Use Management. 5 (2): 303-313.

Nwite, J. C., Essien, B. A., Keke, C. I.,Igwe, C. A. and Wakatsuki, T.(2016).Effect of Different Land Preparation Methods for Sawah System Development on Soil Productivity Improvement and Rice Grain Yield in Inland Valleys of Southeastern Nigeria.Advances in Research/ SCIENCEDOMAIN international, 6(2): 1-17.

Buri M. M. ,Issaka R. N., Wakatsuki, T. and Kawano, N. (2012). Improving the productivity of lowland soils for rice cultivation in Ghana: The role of the 'Sawah' system. Journal of Soil Science and Environmental Management, 3(3): 56-62.

Srisa-ard, K. (2008). Effect of Land Preparation Methods on Growth, Seed Yields of Jasmine 105 Paddy Rice (Oryza sativa L.) and Growth of Weeds, Grown in Northeast Thailand. Pakistan Journal of Biological Sciences, 11(1):34-40.

Sah, G., Shah,S.C., Sah, S.K., Thapa, R.B., McDonald, A., Sidhu, H.S., Gupta, R.K., Sherchan, D.P., Tripathi, B.P., Davare, M. and Yadav, R.(2014).Tillage, crop residue, and nitrogen level effects on soil properties and crop yields under rice-wheat system in the Terai region of Npal. Global Journal of Biology, Agriculture & health Sciences, 3(3):139-147.

7

Estimation of Roof Water Harvesting Potential of Navsari City by Remote Sensing and GIS

D.K. Dwivedi and P.K. Shrivastava

Introduction

In India, the ground water recharge in arid and semi-arid regions is very important particularly as most of the population residing in these regions depend on groundwater for drinking and irrigation purposes. In such scenarios, to alleviate the problems of drinking water, the collection and storage of rainwater to supplement existing water supply sources should be implemented. The utilization of rainwater is one of the best methods that can aid in sustainable urban development (Kim *et al.*, 2005). Every drop of rainfall should be utilized to cater the demand of an ever increasing population as the present system of water supply may not be sufficient in future. The rain water collection & its utilization, seems to be the only option to satisfy the demand of arid and semi arid regions. In most cases, the rain water is the purest water source & contains very low impurities. It is necessary to adopt rainwater harvesting methods in places where the conventional water supply has failed to meet the peak period demand of the people for drinking water. Rain water harvesting is one of the most economical and practical methods to provide supplementary water supplies for drinking purpose. Also, if the drinking water supply by the rain water harvesting structure is sufficient then the supplementary water can be used for domestic purposes. The only cost effective solution to minimize the demand-supply gap by augmenting the existing water supply seems to be rain water harvesting.

Roof water harvesting is a process of collecting runoff occurring due to rainfall from impermeable terrace on top of the house and conserving it in a tank. This method is comparatively easier and reliable and provides good quality water for drinking. The collected rainwater can either be used as the principal or as a supplementary source of water. The stored water could be used for drinking as well as other domestic purposes like cooking, washing and bathing and agriculture purposes like irrigating land, feeding cattle etc.

Roof water harvesting is a proven approach towards conserving water effectively and economically. As rainfall in many regions of India is characterized by heavy rains of short duration and long duration devoid of rainfall,. The roof water harvesting method has been used since ancient times for providing water for drinking during water shortage. The roof water harvesting concept varies from small and basic, such as attachment of a pipe draining into an underground tank, to large and complex, such as those to collect water from many hectares serving a large number of people. Roof water harvesting was used mostly in areas which lacked alternative forms of water supply such as coral islands (Datar *et al.*, 2006; Devi *et al.* 2005) and arid remote locations devoid of suitable surface or groundwater resources (Tripathi and Pandey, 2005). The average annual per capita water availability in 2001 and 2011 was assessed at 1,816 cubic meters and 1,545 cubic metres, respectively. Annual per capita water availability of less than 1,700 cubic meters is considered a water-stressed condition, where as annual per capita water availability below 1,000 cubic meters is considered as a water scarcity condition (INDC report, 2015).

Lade and Oloke (2013) assessed the rain water harvesting potential in Ibadan, Nigeria by hydrological analysis using rainfall data for 30 years from two meteorological stations, with the aim of providing a more sustainable. Rooftop water harvesting was found to be technically feasible based on the prevailing rainfall pattern with over 90% of households having a rooftop constructed from technically appropriate materials. Results of the study indicated that an average roof of 80 m^2collected 82,835 liter/yr (45 liter/ person/day) for a family of five people which was about the required water demand for drinking and cooking purposes. Ganguly *et al.* (2014) studied the prospects and possibilities of rain water harvesting in Shimla region of Himachal Pradesh, which experienced water shortages in summer periods due to increase in population and also climatic conditions. The result of their study showed that rain water can be stored throughout the year without incurring any significant losses. The results showed that the total volume of rain water harvested would meet 30% of the total water demand requirement annually for washing and sanitation purposes. Panigrahi (2017) conducted a study aiming at developing awareness towards judicious use of water among masses and efficient ways to harvest roof top rain water resources at institutional / multi-storied buildings in Malda district of West Bengal. It was concluded that Malda district had a huge potential of rooftop rain water harvesting. Mohanty *et al.* (2018) conducted a study of roof water harvesting structures in Marijita village of Jagatsinghpur district, Odisha.

The total water demand for the selected household was about 1,85,400 liters, out of which 5400 litres of water was used for drinking purpose considering 3 litres of water per person per day, while 1,80,000 liters of water required for other use with 100 liters per person per day. The rooftop water harvesting system provided 1,48, 320 liters of water for the household.

Study area

Navsari city (20.9467° N and 72.9520° E) situated in Navsari district of South Gujarat was selected as the study area. It is the 9th biggest municipal corporation of Gujarat. The city is located at an elevation of 9 m from the mean sea level. According to 2011 census, the population of Navsari city was 1.71 lakhs. The average annual rainfall of Navsari is 1621 mm.

Table 1: Run off coefficients for various roof tops (Pacey and Cullins, 1989)

Type of catchment	Coefficients
Tiles	0.8-0.9
Corrugated metal sheets	0.7-0.9
Concrete	0.6-0.8
Brick pavement	0.5-0.6

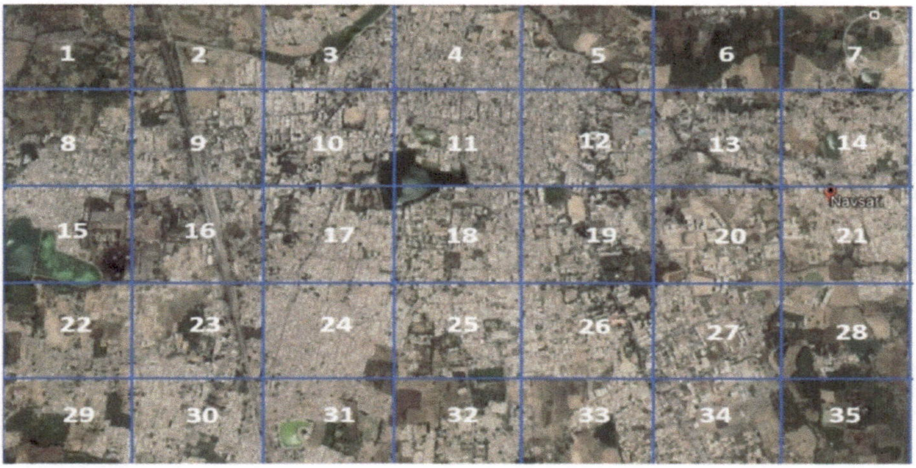

Fig. 1: Satellite image of Navsari city (28/10/2018)

Materials and Methods

In this study, the roof water harvesting potential of Navsari city has been assessed using remote sensing, GIS and ground truthing. The total amount of water received in the form of rainfall over an area is called the rainwater endowment of that area and the amount that can be effectively harvested is called the water harvesting potential.

Area of catchment (m^2) × Rainfall deth (m) = Volume of water received (m^3)

The following information was required to estimate the rainwater harvesting potential.

1. Mean Rainfall depth
2. Catchment area characteristics
3. Run off and its coefficient

Run off depends upon the area and type of the catchment over which it falls as well as surface features. To obtain the volume of the rain water harvested, the depth of rainfall was multiplied with the catchment area. The effective area of roof used for computing volume of water is called roof foot print. For plane horizontal rectangular or other shape roof, roof foot print was same as the roof and thus effective area was same. The run off coefficient is the value that represents the ratio of the volume of water that runs off a surface to the volume of rainfall that falls on the surface There are many parts of the hydrologic cycle that affect the runoff coefficient. Thus run- off coefficient for roof depends on roof type, slope, degree of imperviousness and surface roughness.

The city was divided into 35 grids of equal dimensions and "Remap software" was used for land use/land cover classification of each grid in which classes such as roads, barren land, built-up area, agriculture land, ponds etc were differentiated. Remap uses the power of the Google Earth Engine, allowing directly access of satellite data archives and state-of-the-art remote sensing methods. To achieve the lowest error rate and highest classification accuracy, it was ensured that the training points were sufficient in number (>20) and accurately located for each class.

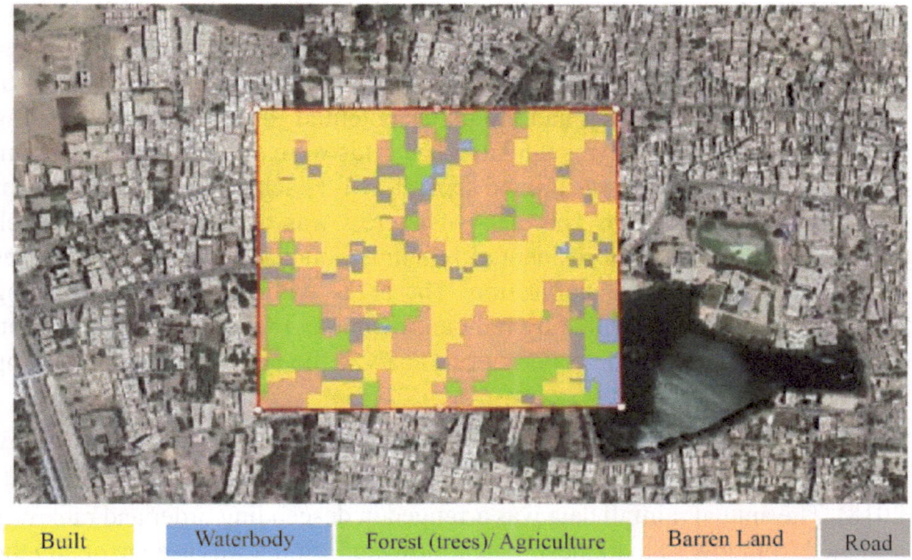

Fig. 2: Land use / land cover classification of Grid 10 of Navsari city

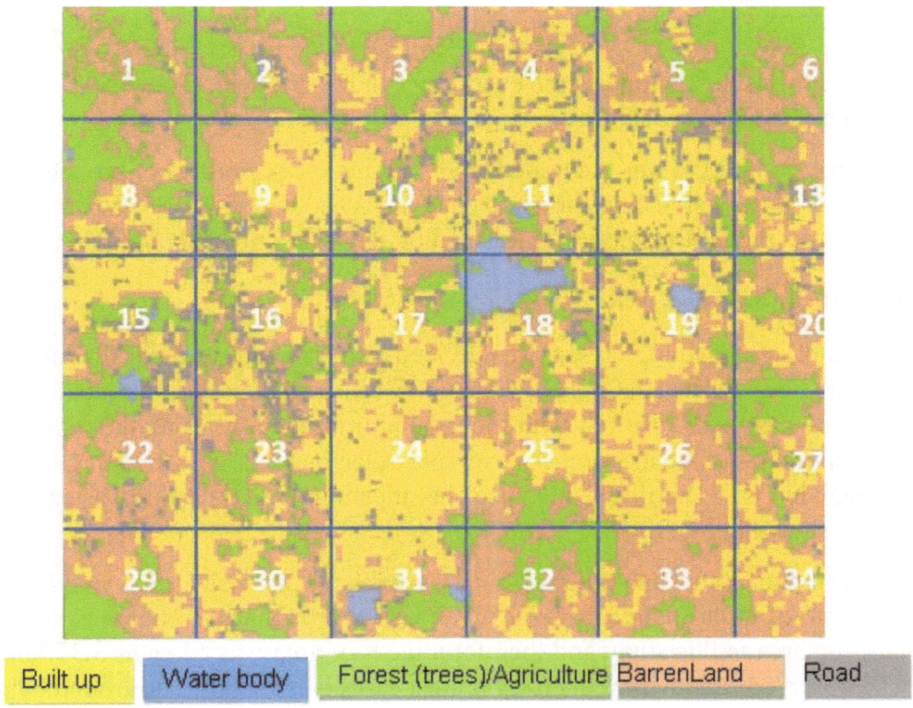

Fig. 3: Land use / Land cover Classification of Navasari city(Grids : 1 to 35)

The city has been continuously expanding due to increasing population and increasing number of houses in the outskirts of the city. The satellite image of October 2018 was used to assess the roof area of the city. The city area of 17.5 km^2 was selected as it had the major potential for rain water harvesting and the area of each grid was 0.5 km^2. The grids were assigned numbers from 1 to 35 and the software was trained separately for each grid to minimize error in land use classification. The class of interest was the built-up area which included commercial and residential buildings, bungalows and small houses with roof area for water harvesting. The areas with roof water harvesting potential were thus identified by remote sensing and GIS. In addition to that, ground truthing was also carried out to analyze the areas capable of rooftop rain water harvesting.

The annual rainfall at various recurrence intervals was obtained using the W eibull's proability plotting formula using the rainfall data of 35 years (1984-2018). The annual rainfall data was arranged in descending order and rank was assigned from lowest to highest.. The following formula was used to obtain the exceedence probability of rainfall.

$$m = \frac{P(x \geq X)}{n+1}$$

Where:' m' is the ranking of the event, 'n' is number of events and 'P (x ≥ X)' is the exceedence probability.

Recurrence interval was then obtained by taking the reciprocal of probability. The roof water harvesting for recurrence intervals of 2, 5, 10, 25, 50 and 100 years were then estimated using the methodology described above.

Results and Discussion

The satellite image of Navsari was divided into 35 equal grids of 0.5 km^2 as shown in Fig-1. The land use classification was carried out for individual grids and they were also checked with the ground truth data by using the latitudes and longitudes of the location. An example of land use classification of Grid 10 is given in Fig. 2. Similarly, land use classification of all the grids was carried out as shown in Fig.3.

The compound area and other adjoining areas have also been accounted for in 5.62 km^2 of built-up area. The roof area was approximated to be 60 % of the built-up area in this study and therefore, the roof area was taken as 3.37 km^2. The observed runoff from the roof in most cases was found to be around 60% of the total rainfall and therefore, the runoff coefficient considered for concrete roof was taken as 0.6. The amount of rainwater that could be obtained from the rooftop was found to be 3280 million liters per year considering average rainfall of 1621 mm. However, it will not be possible to store all the available water obtained from rooftop in all the houses and buildings, as it requires provision for an underground tank in the vicinity or in the basement of the residential area. Assuming that only 5 % of the available roof water could be stored, the estimated roof water harvesting potential was found to be 0.164 MCM.

Table 2: Land use / land cover classification of Navsari city

Sr.No.	Land use / Land cover	Area (km²)	Percentage Cover
1	Water bodies	0.28	2
2	Barren land	6.65	38
3	Roads	1.20	7
4	Trees/ Agriculture	3.75	21
5	Built up	5.62	32
	Total Area	17.50	100

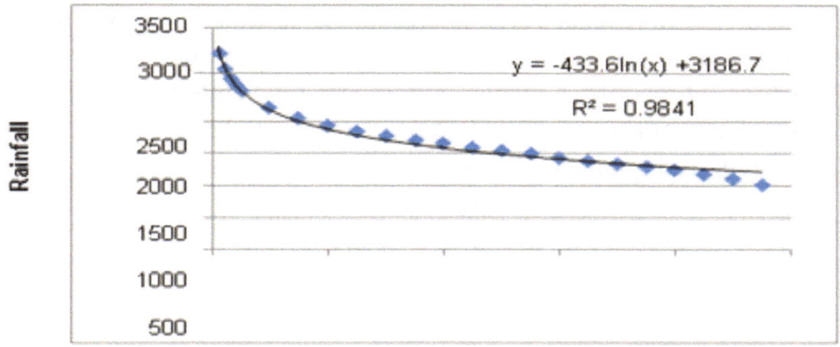

Fig. 4: Plot of Annual Rainfall vs. Probability

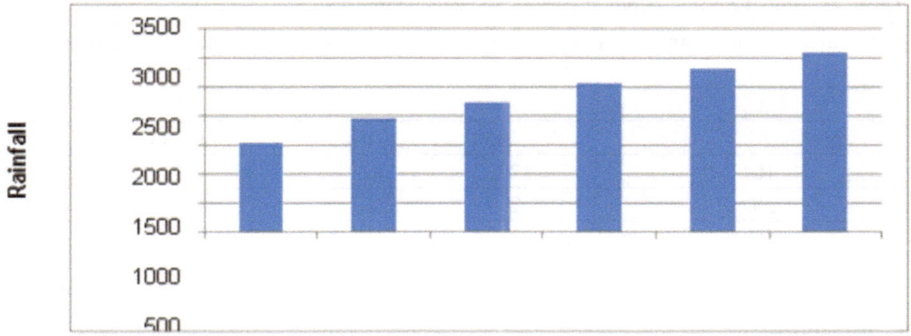

Fig. 5: Annual Rainfall depth (mm) at various recurrence intervals

According to WHO, an average person requires around 4 liters of drinking water per day and therefore approximately 1460 liters of drinking water per year. The estimated amount of 164 million liters for an average annual rainfall is capable of satisfying drinking water requirements of approximately 1.12 lakh people which is around 65.6 % of the total population of Navsari. The additional amount of water could be used for other domestic purposes. However, if the water is to be used for drinking, reliable commercial filter of adequate capacity should be installed along the pipeline or a filter using appropriate media underground before the water is stored in the tank. In case, sufficient space is not available for an underground tank, the pipes carrying water from the roofs of several houses can be connected for diverting water to a nearby underground water tank which can be built in the barren land or common plot. If the annual rain fall is 800 mm which is 50% deficient than average annual rain fall, the estimated roof water harvesting potential would be 0.162 million liters which can satisfy the drinking water requirement of 55,397 people of Navsari city which is around 32.4 % of the total population of Navsari city For return periods of 10 years, 25 years, 50 years and 100 years, the roof water harvesting potentials were estimated to be 0.226, 0.261, 0.287 and 0.312 MCM respectively. Therefore, if sufficient provisions are made to harvest rainfall, the roof water can successfully reduce the drinking water problems of the people, however, significant investment is required to construct tanks of appropriate sizes in the house, residential buildings, schools etc. to store the required quantity of water.

Table 3: Estimated roof water harvesting potential for average rainfall depth

Roof area	**3.37 km^2**
Average Annual Rainfall	1621 mm
Runoff coefficient	0.6
Available roof water	3.280 MCM
Roof water harvesting potential (10%)	0.164 MCM

Table 4: Annual rainfall depth at various recurrence intervals

Return period (years)	**Annual Rainfall (mm)**	**Roof water harvesting potential (MCM)**
2	1544	0.156
5	1958	0.198
10	2232	0.226
25	2578	0.261
50	2835	0.287
100	3090	0.312

Conclusion

Rainwater harvesting through rooftops is an appropriate method of water conservation and Indian government has mounted this as a part of watershed management programme. With the changing climate, high intensity rains with less number of rainy days or uneven distribution of rainfall both spatially and temporally results in instant flooding and runoff. Moreover, due to removal of vegetative cover and denuding forests along with concretisation, rain water doesn't get sufficient opportunity time to recharge groundwater aquifers, which results in water scarcity immediately after monsoon. Therefore, an attempt was to examine the present status of drinking water requirement of Navsari and estimated rooftop rain water potential. The estimated amount of roof water harvesting potential was found to be 164 million liters which is capable of satisfying drinking water requirements of approximately 1.12 lakh people. The additional amount of water can be used for other domestic purposes.

References

Datar, R. (2006), Designing and implementing rainwater harvesting systems for industries – case study from Mumbai, 2nd Joint International Conference on Sustainable Energy and Environment (SEE 2006), 21st – 23rd November, 2006, Bangkok, Thailand.

Devi, B.L., Maheshwari, B. and Simmons, B. (2005), Rainwater harvesting for residential irrigation: how sustainable is it in an urban context, Proc. of 12 International Conference on Rainwater Catchment Systems, New Delhi, India.

Ganguly, R., Bansal, A., Mishra, M., and Kumar, A. (2014). Application of Rain Water Harvesting Scheme in Shimla Region. Hydrology: Current Research, 5(3), 1.

Intended Nationally Determined Contributions Report (2015). Template for Vulnerability Assessment, Mitigation and Adaptation. Government of India. Ministry of Water Resources, River Development and Ganga Rejuvenation. National Water Mission.

Kim, R. H., Lee, S., & Kim, J. O. (2005). Application of a metal membrane for rainwater utilization: filtration characterization and membrane fouling. Desalination, 177, 121–132.

Lade, O., and Oloke, D. (2013). Assessment of rainwater harvesting potential in Ibadan, Nigeria. Environmental Engineering Research, 18(2),91-94.

Mohanty R., Dalai A., Singh K., Das A. and Dalai B. (2018). New approach in rain water harvesting through rooftop water management technique: case study. International Journal of Agriculture Sciences, 10(8), 5815-5816.

Murray, N.J., Keith, D.A., Simpson, D., Wilshire, J.H. & Lucas, R.M. (2017) REMAP: An online remote sensing application for land cover classification and monitoring. Bio Rxiv. DOI: 10.1101/212464

Murray, N.J., Keith, D.A., Simpson, D., Wilshire, J.H., Lucas, R.M. (2017) REMAP: The remote sensing ecosystem monitoring and assessment pipeline. https://remap-app.org

Pacey, A. and A. Cullins.(1999). Rain water harvesting. The collection of Rainfall and Runoff in the Rural areas. Technical centre for Agricultural and Rural cooperation (ACP – EEC Lome Convention). 216.

Panigrahi (2017) Roof-Top rainwater harvesting system for official / multistoried building with reference to Malda district, West Bengal. International Journal of Engineering Research and Application, 7(1), 111-115.

Tripathi, A. K. and U. K. Pandey. (2005), Study of rainwater harvesting potential of Zura village of Kutch District of Gujarat, Journal of Human Ecology 18(1), 63-67.

World Health Organization, WHO/UNICEF Joint Water Supply, & Sanitation Monitoring Programme. (2015). Progress on sanitation and drinking water: 2015 update and MDG assessment. World Health Organization

8

Dry Synthesis and Characterization of Molybdenum Trioxide Nanoparticles for Seed Treatment of Pulses

S.K. Dwibedi, J. Nanda, P.K. Samant, B. Behera, S. Jena, M. Behera and R.K. Dwibedi

Introduction

Pulses are a part of healthy and balanced diet and have special importance for vegetarians in fulfilling their daily requirements for body proteins. Low productivity in pulses is due to nutritional deficiency in soil and imbalanced external fertilization (Awomi *et al.*, 2012). Nitrogen (N) is more important as it is essential component of amino acids and proteins. It fixes atmospheric N of about 35 kg ha^{-1} by the help of symbiotic *Rhizobium*, a soil bacterium (Movalia *et al.*, 2020). Globally, *Rhizobium* and *Bradyrhizobium* can fix approximately 175 x 10^6tons of N every year which is higher than the annual total chemical N fertilizer production of 30 x 10^6tons (Hidayatullah *et al.*, 2016). The element molybdenum (Mo) is crucial for functions of nitrogenase and nitrate reductase, the two important enzymes that act as catalyst for N fixation. These two enzymes work properly in presence of Mo as nitrogenase contains Fe-Mo and Fe-Fd proteins, and N fixation cannot take place without these proteins (Hidayatullah *et al.*, 2016). Mo acts as cofactor of the proteins required for electron transfer in the synthesis of nitrogenase complex which is responsible for conversion of N_2 to ammonium (Glowackaet et al., 2019). Apart from N metabolism, it also helps in O and S cycles. Plants get Mo from soils and can assimilate in Molybdate. Sandy and acidic soils contain less available Mo. In low pH, the MoO_4 adsorbs onto positively charged metal oxides (Fe, Al, and Mn) and the maximum adsorption occurs between pH 4 to 5 (Smith *et al.*, 1997).

With every unit increase in pH above 3, the $MoO4^-$ availability increases 100 folds. Thus application of lime to adjust soil pH is essential for soluble Molybdate. However, soluble Mo can also form ionic compounds with Na, K,

Ca and Mg and can also form complexes mostly with humic and fulvic acid (Hidayatullah *et al.*, 2016; Kaiser etal., 2005). Ill drained soils, such as peat and marshy organic rich soils tend to accumulate high MoO4 levels resulting in molybdenosis in ruminants (Gupta, 1997) whereas in well drained sandy soil Mo is leached down (Kaiser *et al.*, 2005). Being a trace or minor element, Mo is required in a very small quantity by plants but continuous cropping in well drained light textured acid soils show deficiency symptoms in the succeeding secondary crops such as crucifers, legumes and citrus (Grant, 2018).

Soil application of Mo may require large quantities depending on the crop, soil type and inherent soil contribution. It is subjected to losses through percolation, leaching, runoff and weed uptake. Mo may also get fixed with the oxides of other minerals and become unavailable to the current season crop. Increase in soil pH by liming is although a feasible option but it may add to the pool of greenhouse gases due to release of carbon dioxide (West and McBride, 2005). Foliar application is another viable option but is practicable only after development of crop canopy and till then the crop may suffer from irreparable loss. Foliar application is prone to weather parameters (Hidayatullah *et al.*, 2016).

One more option for supplementing crop requirement for Mo is through seed pelleting and seed dressing but it may impair seed respiration, reduce survival of rhizobia, plant nodulation and efficiency of N2 fixation (Almeida et al., 2013). The proportion of applied Mo that enters into the seeds and actively plays role in plant metabolism is also not so adequate. However, no other alternative mode of application of Mo such as use of nanoparticles of Mo to avert the above mentioned negative effects of soil, foliar and seed application are traceable. Nanoparticles, due to their unique physicochemical characteristics with ultrafine particle size of 1-100 nanometers (nm), high mobility and surface to volume ratio have the potential to modulate the redox status (European Commission Recommendation of 18 October 2011) and affect the germination, growth, quality and performance of the plants. These materials easily enter into plants by overcoming all barriers at cellular level (Duhanet *et al.,* 2017).

Nano form of molybdenum oxide (MoO3) could be synthesized in laboratory by pulsated laser ablation and hydration method (Sobanska *et al.*, 2020) and nanosecond laser ablation method (Romeo, *et al.*, 2019). Manivel and Lee (2015) synthesised one-dimensional MoO3 by thermal, microwave and sonochemical methods. Microwave assisted synthesis of MoO3 nanoparticles by Tari *et al.* (2015) produced particles of 50 nm as characterized by XRD, TEM and SEM. Molybdenum oxide nanoparticles were also synthesized by using sol-gel method (Ganguly and George, 2007). The yellow powder thus formed was characterized in SEM and XRD had 10-100 nm size. Wongkura *et al.* (2013) could synthesise MoO3 nanoparticles by combination of refluxing and calcination. Nano crystalline powders of MoO3 were synthesized from ammonium heptamolybdate tetrahydrate as initial material by using precipitation method at various pH values that resulted in platelet-like shape with size ranging from 73 to 760 nm depending on pH (Jittiarporn *et al.*, 2016). Modified co-precipitation method was used to synthesize MoO3 by using polyethylene glycol as stabilizer in aqueous medium. The average grain size was estimated to be 6.82 nm (Shelke *et al.*, 2019). MoO3 nano cubes were synthesized by Muthamizh *et al.* (2015) by simply heating analytical grade ammonium molybdate in muffle furnace at 600 ^{0}C for 5 hours. The XRD, SEM and TEM analysis showed cube like crystalline cubes. Green synthesis of nanoparticles of MoO3 from *Syzygium aromaticum* flower was successfully done by Kanneganti *et al.* (2014) and the particles were found to be spherical and 19.47 nm size. Molybdenum disulfide nanoparticles were synthesized by wet chemical process that led to formation of nano flowers and nano sheets within range of 10 to 50 nm (Vattikuti and Byon, 2015). Wet chemical method was also followed by Chitturi and Deepak (2014) which produced MoS2 nano particles of 112 nm. Bioinspired synthesis of Mo nanoparticles by using fungus *Aspergillus tubingenesis* TFR29 and their subsequent characterization by TEM, SEM, and AFM indicated size between 2-7 nm (Thomas *et al.*, 2017).

All the above methods are wet chemical methods that require costly equipments, longer period for synthesis and are beyond the reach of the common farmers due to higher market price. Hence, a simpler and cheaper method for synthesis of nano molybdenum trioxide was highly essential for seed treatment of pulses that may improve efficacy of nitrogen assimilation.

Materials and Methods

The nanoparticles of Mo O3 were synthesized taking commercially available ammonium orthomolybdate {(NH4)2 Mo O4} in silica crucible and calcinated at 600 °C in a muffle furnace for 5 hours. The Mo O3 nanoparticles thus formed were kept inside airtight zipper polythene bag and then stored inside refrigerator at 4 ^{0}C to prevent agglomeration. The nano particles were then sent to the Central Institute of Plastics and Engineering Technology, Bhubaneswar (CIPET) for characterization by Transmission Electron Microscope (TEM) and Power X-ray diffraction (XRD) analyses.

a) Transmission Electron Microscope(TEM)

TEM uses an electron beam to measure size and distribution of nanoparticles, providing much higher resolution than with light-based imaging techniques

b) X-Ray diffraction(XRD)

From the XRD graph, average crystallite size in the MgO nanoparticle was calculated by following Debye-Scherrer formula as:

$$D = \frac{k\lambda}{\beta \cos\theta}$$

Where: D - Crystal size, k - Shape factor (usually 0.9), λ - Wavelength of X-ray source used β - Line width at half maximum and θ - Bragg diffraction angle

Results and discussion

a) Transmission Electron Microscope (TEM) analysis

The Fig. 1 illustrates TEM image of MoO_3-NP. From the image it is observed that a small amount of agglomeration is present in the synthesized ash-sample. The MoO_3 particles are nearly monodisperse oblong shaped crystallines with the average size seen in the micrograph is below 100 nm at least in one dimension. Upon calcinations of Ammonium molybdenum, the particles were transformed into orthorhombic (α) MoO_3. Presence of small amount of agglomeration in the submitted sample as observed from the TEM image could possibility be due to absorbance of moisture during storage and characterization.

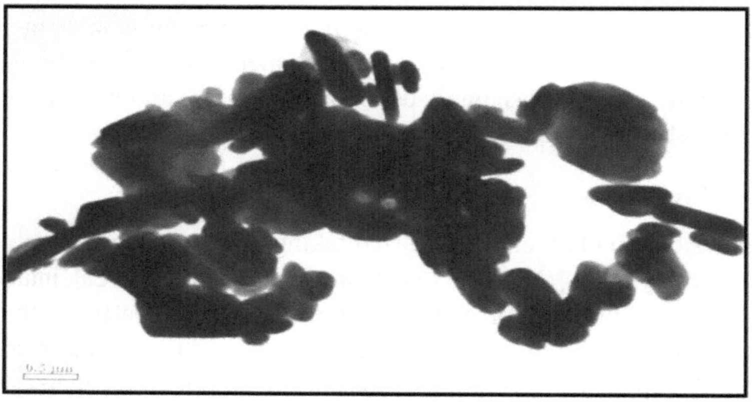

Fig. 1: TEM image of Molybdenum trioxide nanoparticles

b) X-ray diffraction (XRD)analysis

The XRD pattern of the MoO_3 sample, calcined at 600 °C for 5 hrs, is presented in Figure 2.

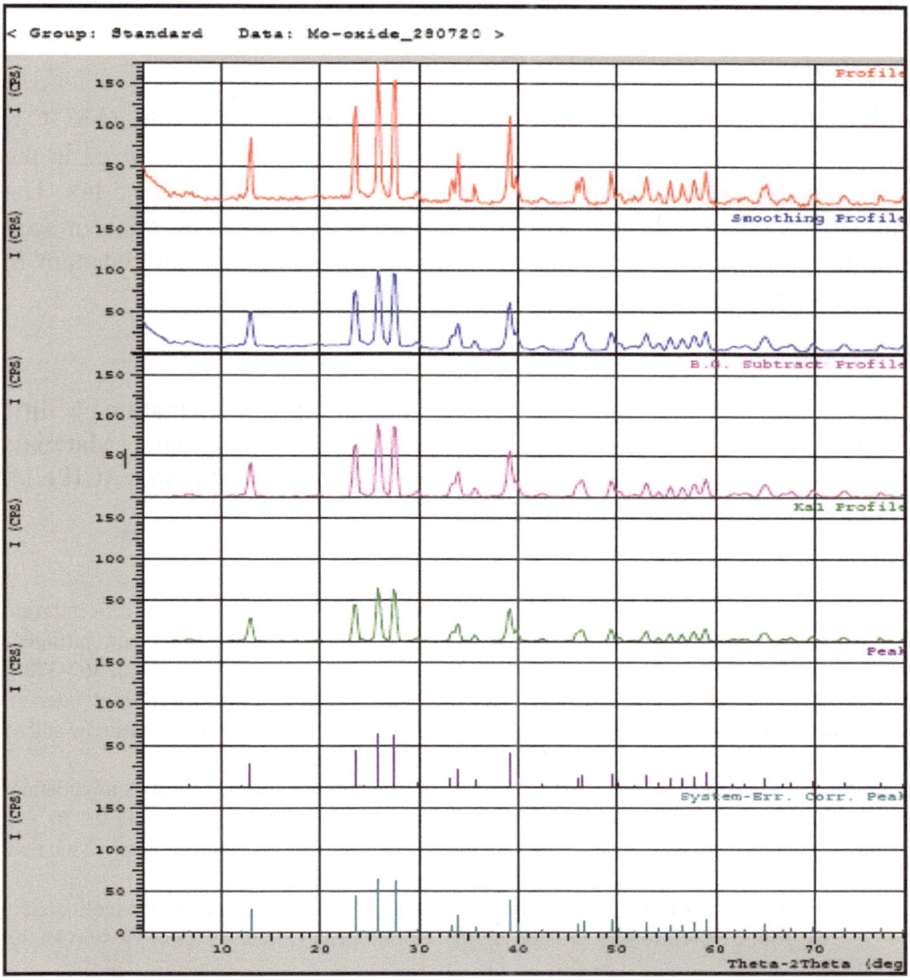

Fig. 2: X-ray diffraction (XRD) analysis of Molybdenum trioxide nanoparticles

The average particle size in the calcined MoO_3 nanoparticles was estimated by the Scherrer equation (Dinesh, 2012).

$$D = \frac{k\lambda}{\beta cos\theta}$$

Where, D is the crystallite size, k is the shape factor, which usually takes a value of about 0.89, λ is the wavelength of X-ray source used (0.15406 nm for Cu Kα), β is the full width at half-maximum (FWHM) in radians (0.00885231), and θ is the Bragg diffraction angle in radians (0.2259381077). The average particle size of MoO3, calcined at 600 °C for 5 hrs, was measured to be 68.55 nm. Both XRD and TEM analyses of samples were in concurrence with each other.

Conclusion

Oblong shaped nanoparticles (NPs) of Molybdenum trioxide (Mo O3) with average particle size of 68.55 nm were successfully synthesized in the laboratory by calcination of Ammonium molybdate at 600 ^{0}C for 5 hrs. The MoO3-NPs thus synthesized could be further used as seed dressing or seed priming of pulses so as to improve the efficacy of treated molybdenum in fixation of atmospheric nitrogen mediated by *Rhizobium* spp

Acknowledgement

The technical and informative supports of the Central Instrumentation Facility, OUAT and Laboratory for Advanced Research of Polymeric Materials (LARPM), Central Institute of Plastics Engineering & Technology (CIPET), Bhubaneswar are highly acknowledged.

References

Almeida FFD, Araujo AP and Alves BJR. 2013. Seeds with high molybdenum concentration improved growth and nitrogen acquisition of rhizobium-inoculated and nitrogen-fertilized common bean plants. Revista Brasileira de Ciência do Solo, 37(2): 367-378.

Awomi TA, Singh AK, Kumar M. and Bordoli LJ. 2012. Effect of phosphorous, molybdenum and cobalt nutrition on yield and quality of mungbean (Vigna radiata L.) in acidic soil of Northeast India. Indian Journal of Hill Farming, 25(2): 22-26.

Chitturi V and Deepak K. 2014. Synthesis and characterization of molybdenum disulfidenanopowder. International Journal of Research & Technology, 2(4): 27-30.

Duhan JS, Kumar R, Kumar N, Kaur P and Nehra K. 2017. Nanotechnology: The new perspective in precision agriculture. Biotechnology Reports, 15: 11-23.

European Commission Recommendation of 18 October 2011 on the definition of nanomaterials (text with EIA relevance). Official Journal of European Union, L 225, p-38-40 (Oct 20, 2011).

Ganguly A and George R. 2007. Synthesis, characterization and gas sensitivity of MoO3 nanoparticles. Bulletin of Material Science, 30(2): 183-185.

Glowacka A, Gruszecki T, Szostak B and Michalck S. 2019. The response of common bean to sulphur andmolybdenum fertilization. International Journal of Agronomy. <http:// downloads. hindawi. com / journals/ija/2019/3830712.pdf>. Grant BL 2018. What is molybdenum: Information on molybdenum sources for plants. <https:// www. gardeningknowhow.com/garden-how-to/soil-fertilizers/what-is- molybdenum.htm>

Gupta UC. 1997. Soil and plant factors affecting molybdenum uptake by plants. In: Gupta UC, ed. Molybdenum in Agriculture. Cambridge University Press.

Hidayatullah, Mahammad T, Ahmed SK, Shah S G and Mahammad U. 2016. Response of peas to molybdenum application with and without Rhizobium inoculums under alkaline calcareous soils. Science International (Lahore), 28(5):4753-4758.

Jittiarporn P, Sikong L, Kooptanord K, Taweepreda W, Chooto P and Khangkhamano M. 2016. Synthesis of h- MoO3 and (NH4)2Mo4O13 using precipitation method at various pH values and their photochromatic properties, Applied Mechanics and Materials, 835: 34-41.

Kaiser BN, Gridley KL, Brady JN, Phillips T and Tyerman SD. 2005. The role of molybdenum in agricultural plant production, Annals of Botany, 96: 745-754.

Kanneganti A, Kumar KA and Vivek V. 2014. Green synthesis of molybdenum trioxide nanoparticles using Syzygium aromaticum flower buds extract. International Journal of Scientific Research and Development, 2(4).

Manivel A and Lee GJ. 2015. Synthesis of MoO3 nanoparticles for azo dye degradation by catalytic ozonation. Materials Research Bulletin, 62: 184-191.

Movalia J, Donga S and Parmar KB. 2020. Effect of boron and molybdenum on summer green gram (Vigna radiata L.) (GM-4) under medium black calcareous soils: A review. Proceeding of the National Conference on Innovations in Biological Sciences (NCIBS), January 10, 2020.

Muthamizh S, ZSuresh R, Giribabu K, Manigandhan R, Praveen Kumar S, Munuswamy S, Stephen A Narayanan V. 2015. Molybdenum oxide Nanocubes: Synthesis and characterizations. AIP Conference Proceedings; ISSN 0094-243X.

Romeo NOEZ, Miguel A, Camacho-Lopez Alfredo R, Vilchis-Nestor VH, Castrejon-Sanchez, Guillermo Aguilar, Santiago Camacho-Lopez, Marco Camacho Lopez. 2019. Synthesis of molybdenum oxide nanoparticles by nanosecond laser ablation. Materials Chemistry and Physics, doi: https:// doi.org/10.1016/j.matchemphys.2019.122163.

Shelke PD, Gawali DB and Rajbhoj AS. 2019. Synthesis and evaluation of antioxidant behavior of MoO3 nanoparticles. International Journal of Research and Analytical Reviews, 6(1): 1616-1620.

Smith KS, Balistriri LS, Smith SS and Severson RC. 1997. Distribution and mobility of molybdenum in terrestrial environment. In: Gupta UC, ed. Molydbedum in Agriculture. Casmbridge University Press.

Sobanska Z, Zapor L, Szparaga M and Stepnik M. 2020. Biological effects of molybdenum compounds in nanosized forms under in vitro and in vivo conditions. International Journal of Occupational Medicine and Environmental Health, 33(1): 1-19.

Tari F, Manteghain M and Bozorgi B. 2015. Microwave-assisted synthesis of molybdenum oxide nanoparticles. Particle science Technology, 121-127.

Thomas E, Rathore I and Tarafdar JC. 2017. Bioinspired production of molybdenum nanoparticles and its effects on chickpea (Cicer aerietinum L.). Journal of Bionanoscience, 11: 1-7.

Vattikuti SVP and Byon C. 2015. Synthesis and characterization of molybdenum disulfide nanoflowers and nanosheets: Nanotribiology. Journal of Nanomatrials.

West TO and McBride AC. 2005. The contribution of agricultural lime to carbon dioxide emissions in the United States: Dissolution, transport and net emissions. Agriculture Ecosystem and Environment, 108: 145-154.

Wongkura P, Tongtem T and Thongtem S. 2013. Synthesis of h- and α- MoO3 by refluxing and calcinations combination: Phase and morphology transformation, photocatalysis and photosensitization. Journal of Nanomaterials, doi: http://dx.doi.org/10.1155/2013/702679.

9

Protected Cultivation Green House and Hydroponics Technology

N. Yadav, K. Barman, P. Belwal, A. Kanujiya and B.K. Dadrwal

Introduction

Protected cultivation or greenhouse cultivation is the most common technique to grow vegetable crops qualitatively and quantitatively, and has spread widely all over the world in recent decades. Greenhouse agriculture is also known as environmentally controlled planting. The technology involves the cultivation of vegetable crops in a regulated environment in which factors such as temperature, soil, light, humidity, irrigation, fertilizers, etc. are controlled in order to achieve optimum production and ensure a steady supply even in off-season conditions. In India, the total area under protected cultivation or green house is currently approximately 25 000 ha whereas the protected vegetable area is approximately 2000 ha which has been challenged by land ownership limitations, decreasing crop production, rapid urbanization, ever increasing population and decreasing biodiversity. Food demand, especially vegetables, has increased growing of different crop varieties as protected crops, or greenhouse has opened up a new dimension to extra production within a limited crop area.

Designs, construction, operations, location and site of Greenhouse

Protected vegetable cultivation, being a concentrated activity, involves perfect planning and many stages of operation for its realization. The design of a greenhouse varies depending on its location, be it in a temperate, tropical or desert environment. Glasshouses are more preferred in the country's cold or temperate regions than normal greenhouse but greenhouses provide 'windbreak effect' and 'shading effect' in other tropical and sub-tropical regions. During the rainy season rain shelters are the traditional protective structures to prevent flooding. In the dry regions, however, the temperature and humidity inside the covered house provides a 'oasis effect' compared to the heat and dryness outside the greenhouse area. Greenhouses with the light intensity should be away from overpopulated areas, urban and level fields. If the greenhouse has to be built on a slope, precaution should be taken so that it is directed away from the surface runoff. Satisfactory power source and water supply have to be accessible closer to the greenhouse.

Types and materials for greenhouse

Greenhouses for vegetable cultivation include polyhouse, glasshouse, polyhouse with low tunnels, net insect-proof house, polyhouse with zero energy etc. There are various types of greenhouse structures such as closed polyhouse with fan and cooling-pad system, open ventilated, shade net house including sloping roof, rain shelter, etc. Greenhouse structures are constructions based on a variety of shapes (ridge and furrow type, lean to type, evan span type, etc.), utility (controlled temperature and humidity), construction (framed pipe or truss, wood), housing material (plastic film, glass, fiberglass). Plastic film is easy to use as covering materials and different types such as fiberglass reinforced polyester, acrylic, polycarbonate, polyvinyl chloride film and polyethylene film are currently available for use.

Production technology for protected cultivation

In greenhouses, vegetables can be produced under various methods which include planting in soil directly or in soilless cultivation like hydroponic cultivation which involves different soilless mixtures using liquid nutrient, media and containers.

Soil and its management:

It is a conventional method in which vegetable seeds are sown or seedlings are planted directly in well-drained soil. Due to continuous cropping of same or similar crops, soil-borne diseases and pests become a major concern for greenhouse cultivation. Soil preparation involves initial practices like fumigation, sanitation, solarization, mulching, etc. in which solarization commonly adopted for reduce soil-borne pathogens.

Production of vegetable seedlings: In greenhouse, production of disease-free, vigorous and off season seedlings requires proper media sterilization, rapid growth through hormonal treatment, proper fertigation and management for virus and pest free seedling nursery. Currently the plug tray nursery raising technology is used with sterilized, soilless growing media such as vermiculite, per liter and coco-peat in the ratio of 3:1:1. Also one can use growing media with addition of biological agents, hydrogel and organic amendments which assist in raising higher population of vegetable seedlings thereby avoiding soil- borne diseases.

Soil-less cultivation: Plants are grown in different growth media other than soil using peat moss, vermiculite, coir, rock wool, per liter sand, sawdust or nutrient mist solution. It is expensive and intensively followed by developing countries. In case of nutrient recycle, all nutrients provided to plants are retained as compared to soil, where only nitrogen can be recovered.

Crop production in greenhouse: Several vegetable crops such as tomato, capsicum, cucumber, lettuce, celery, spinach, cabbage, cauliflower, beans, pea and coriander are grown in green houses. Crop selection is based on their requirement of climatic conditions, off season demand, distance of market etc. There are many cultural practices followed in protected cultivation or greenhouse such as raising of seedling nursery, bed preparation for transplanting, application of fertilizer, laying of drip line for irrigation, use of different types of mulching materials, transplanting of seedlings, fertigation, pruning, training, pollination management, lowering of plants, deleafing, earthing up, application of growth hormones, pesticides application and harvesting.

References

Albright L D. (2002). Controlling greenhouse environments. Acta Horticulturae. 578: 47–54.

Brentlinger D. J. (2007). New trends in hydroponic crop production in US. Acta Horticulturae. 742: 31–3.

Farrag E S H and Fotouh Y O. (2010). Solarization as a method for producing fungal-free container soil and controlling wilt and root-rot diseases on cucumber plants under greenhouse conditions. Archives of Phytopathology and Plant Protection. 43(4/6): 519–26.

Jensen M H. (2002). Controlled environment agriculture in deserts tropics and temperate regions – A world review. Acta Horticulturae. 578: 19–25.

Nichols M. (2007). Good Agricultural Practices (GAP) and greenhouse crops. Acta Horticulturae. 742:135–8.

Sabir, N., & Singh, B. (2013). Protected cultivation of vegetables in global arena: A review. Indian Journal of Agricultural Sciences, 83(2), 123-135.

Singh B and Sirohi N P S (2008) Protected cultivation of vegetables in India: problems and future prospects. Acta Horticulturae.710: 339-342.

Wittwer, S. H., & Honma, S. (1979). Greenhouse tomatoes, lettuce and cucumbers. Michigan State University Press.

10

Growth and Yield of Black Scented Direct Seeded Rice (*Oryza sativa L.*) as Influenced by Different Nitrogen Levels

K.S. Giri

Introduction

Rice plays a unique role in providing food to the majority of the world population. Rice is the most widely consumed food for nearly three-fourths of the population in India. Now a days scented rice is getting higher market price than normal rice varieties. Black Scented rice is commonly considered as forbidden rice. Different researches proved that black Scented rice is one of the "super food" of the world. Black Scented rice is consumed in various ways like puddings, pulao, bread, etc in different festivals and during important ceremonies. Mostly this crop is prone to lodging but resistance to infestation by insect and pest. Works done in India and abroad indicates that the particular method of establishment, adequate level of nitrogen and a particular time of application of nitrogenous fertilizer can enhance productivity. Nitrogen is the main nutrient that determines rice yield, due to its role in the photosynthesis, biomass accumulation and, spikelet formation (Hasegawa et al., 1994; Yoshida *et al.*, 2006). Direct seeded rice (DSR) technique of establishment in rice is the only spectacular option to cut down the wastage of water. Direct seeding of rice offers certain benefits like, reduced usage of irrigation water, labour, energy, time, reduced emission of greenhouse-gases and better growth of succeeding crops over the transplanting method of establishment.

Materials and Methods

The investigation was carried out at research farm of College of Agriculture, Central Agriculture University, Imphal during the kharif season of 2018-19. The treatments comprised of two establishment methods (broadcasting and line sowing), three levels of nitrogen (40, 60 and 80 kg N/ ha) and two time of nitrogen application(1/3rd at basal + 1/3rd at active tillering+ 1/3rdat Panicle initiation and 1/3rd at 3-4 leaf stage + 1/3rd at active tillering+ 1/3rd at panicle initiation) which were laid out in a Factorial Randomized Block Design (FRBD) with twelve treatment combinations and replicated thrice. Chak-hao (Poireiton) was the test variety of black scented rice used in the experiment. The experimental data obtained were subjected to statistical analysis with the help of analysis of variance technique for testing the significance of the treatment effects and interpretation of the result. The significant of the different source of variation was tested with the help of the F test at 5% level of probability.

Results And Discussion

The observations (Table 2)on plant height, no. of tillers/ m^2, Leaf area index, grain yield, straw yield and harvest index of black scented rice showed that line sowing (T^2) could produced significantly higher values than broadcasting (T^1) except plant height. Similar findings were observed by Kumar and Ladha (2011). In case of levels of nitrogen N3 (80 kg N/ ha) shows significantly higher values in all the parameter taken in to consideration. The result is in conformity with the findings of Sahaet *et.al.* (2017), Singh *et al.* (2015) and Uddin *et a*l.(2013). Similarly in case of time of nitrogen application S^1 (1/3rd at basal, 1/3rd at active tillering and 1/3rd at Panicle initiation) produced significantly superior no. of tillers/ m^2 , grain yield and straw yield than S^2. Among treatment combinations T^2N^3S1 produced significantly highest plant height, no. of tillers/ m^2, Leaf area index, grain yield, straw yield and harvest index than all other treatment combinations except $T^2N^3S^2$. The result was in accordance with the earlier finding of Tayefe *et al.*(2011).

Table 1: Treatment details

Establishment Methods	Notation
1.Broadcast	T1
2. Line sowing	T2
Levels of Nitrogen	Notation
1. 40 kg/ha	N1
2.60kg/ha	N2
3.80kg/ha	N3
Time of Nitrogen Application	Notation
1. 1/3rd at basal, 1/3rd at active tillering and 1/3rd at Panicle initiation	S1
2. 1/3rd at 3-4 leaf stage, 1/3rd at active tillering and 1/3rd at Panicle initiation	S2

Table 2: Effect of establishment methods, levels of nitrogen and time of application of on plant height, no. of tillers/ m^2, Leaf area index, grain yield, straw yield and harvest index of black scented rice

Treatments	**Plant height (cm)**	**No. of tillers/ m^2**	**Leaf**			
			area index (%)	**Grain Yield (kg/ha)**	**Straw Yield (kg/ ha)**	**Harvest Index Y-%**
T(Estab.Methods)						
T1	135.39	437.72	5.34	1886	2918.44	39.14
T2	139.34	458.07	5.42	2022.72	2996.28	40.25
CD (P=0.05)	NS	9.34	0.05	20.44	19.22	0.27
N (levels of N)						
N1	131.67	390.42	5	1745.17	2825.92	38.15
N2	136.4	445.92	5.36	1934.25	2956.5	39.54
N3	144.04	507.36	5.78	2183.67	3089.67	41.41
CD (P=0.05)	5.27	11.6	0.07	25.03	23.54	0.33
S (Time of appl.)						
S1	138.21	459.96	5.4	1990.22	3000.33	39.79
S2	136.52	435.83	5.36	1918.5	2914.39	39.6
CD (P=0.05)	NS	9	NS	19.34	18.2	NS
T2N3S1	151.9	527.43	5.86	2252.67	3157.67	41.63
T2N3S2	144.51	510.67	5.77	2198	3104	41.46
S.E d(±)	5.08	11.27	0.06	24.14	22.7	0.32
CD (P=0.05)	10.54	23.38	0.13	50.06	47.09	0.65

Conclusion

It is clear from the results that line sowing + 80kg N/ha +1/3rd at basal, 1/3rd at active tillering and 1/3rd at panicle initiation stage ($T^2N^3S^1$) achieved higher values with regards to both plant growth characters (plant height, no. of tillers and leaf area index) and yield (growth yield, straw yield and harvest index). So that the treatment combination $T^2N^3S^1$is recommended.

References

Hasegawa T, Koroda Y, Seligman NAG and Horie T. (1994) Response of spikelet number to plant nitrogen concentration and dry weight in paddy rice. Agronomy Journal; 86(4):673-676.

Kumar V and Ladha J K. Direct seeding of rice: recent development and future research needs. 2011; Advances in Agronomy; 111(1):297-413.

Saha B, Panda P, Patra P S, Panda R, Kundu A, Roy A S. (2017) Effect of Different levels of Nitrogen on Growth and Yield of Rice (*Oryza sativa* L.) Cultivars under Terai-agro Climatic Situation. International Journal of Current Microbial Application Science; 6(7): 2408-2418.

Singh D K, Pandey P C, Priyanker Q A and Gupta S. (2015). Nitrogen management strategies for direct seeded aerobic rice (*Oryzasativa* L.) grown in mollisols of Uttarakhand (India). International Journal of Applied and Pure Science in Agriculture 1(7): 2394-5532.

Tayefe M, Gerayzade A, Amiri E and Zade AN. (2011). Effect of nitrogen fertilizer on nitrogen uptake, nitrogen use efficiency of rice. International Conference on Biology, Environment and Chemistry; 24(11): 470-473.

Uddin S, Sarkar MAR and Rahman MM. (2013). Effect of nitrogen and potassium on yield of dry direct seeded rice cv. NERICA 1 in aus season. International Journal of Agronomy and Plant Production; 4(1): 69-75.

Yoshida H, Horie T, and Shiraiwa T. (2006). A model explaining genotypic and environmental variation of rice spikelet number per unit area measured by cross-locational experiments in Asia. Field Crops Research; 97(2-3): 337-343

11

Evaluation and Performance of DSSAT v4.6 For Growth, Development and Productivity of *Kharif* Rice in Southern Bihar

K. Sar, M.R. Chowdhury and S. Dash

Rice is the longest continuously grown cereal crop in the world and according to the International Rice Research Institute (IRRI) it is "one of the most important developments in history" (Forssell, 2008). India has to produce 114 million tons of rice by the year 2030 to meet the food grain requirement of burgeoning population. In state Bihar, *Kharif* is the predominant season with rice cultivated in 3.2 m ha. On the other hand the rice is cultivated in about 1 lakh ha during Rabi season (Anonymous, 2014). The process-based dynamic simulation crop models based on soil, crop and weather factors could be effective research tools for planning alternative strategies for crop management, land use and water management and also a useful tool for planning and developing technological interventions in diverse areas like India (Aggarwal and Kalra, 1994).

As of model (DSSAT v4.6) requirement the weather data (daily basis) on maximum and minimum temperatures, rainfall, solar radiation and relative humidity of six years (2008-2014) for centre Patna were obtained from National Data Centre, Indian Meteorological Department, Pune. Layer wise data on soil parameters such as texture, organic carbon, cation exchange capacity, water-holding characteristics, bulk density, soil pH were obtained from National Bureau of Soil Survey & Land Use Planning (NBSS & LUP),Nagpur. Six years (2008-2014) experimental data conducted at the experiment stations, Patna centre (for rice) of the Bihar Agricultural University, Sabour, Bhagalpur, Bihar, India was used for the calibration and validation of model.

These experiments were carried out to assess the effect of different amounts of nitrogen fertilizer on yield and yield components of rice under irrigated conditions. Treatments included 6 levels of nitrogen fertilizer (0, 40, 60, 80, 120 and 150 kg ha^{-1}) and the experiments were conducted based on a randomized complete block design with three replications. Planting dates was mid July (18-27) for rice using Rajendra Mahsuri-I. Some of the measured factors such as, grain yield, biological yield, leaf area index, plant height, 100 seed weight, days to anthesis and days to maturity, harvest index and N content in straw and grain were provided for the model as observed data for the calibration and validation of model.

Several criteria were used to quantify the differences between observed and simulated data. Test criteria have been separated into two groups, called summary measures and difference measures. Summary measures include the mean of observed values (Obs) and simulated values (Sim), the standard deviation of observations (So) and the simulation (Ss), the slope (b) and intercept (a) of the least square regression (Simi = a + b + Obsi).

The summary measures describe the quality of simulation while the difference measures try to locate and quantify errors. The latter include the mean absolute error (MAE) and the root mean square error (RMSE). They were calculated according to Willmott (1982) as follows and were based on the terms (Simi – Obsi)

$$\text{MAE} = \sqrt{\frac{\sum_{i=1}^{n}(1X_{\text{obs}_i} - 1X_{\text{sim}_i})}{n}}$$

$$\text{RMSE} = \sqrt{\frac{\sum_{i=1}^{n}(X_{\text{obs}_i} - X_{\text{sim}_i})^2}{n}}$$

MAE and RMSE indicate the magnitude of the average error, but provide no information on the relative size of the average difference between (Sim) and (Obs). Varshneya (1999), who gave a simple indication of error in prediction, defined the percent Error (PE). PE is defined as ratio of RMSE to mean observed value expressed as percentage and was calculated as follows:

$$\text{Error Percent} = \sqrt{\frac{(X_{\text{sim}_i} - X_{\text{obs}_i})}{X_{\text{sim}_i}}} \times 100$$

Note: In all equations, X_{obsi} is observed values and X_{simi} is modelled (simulated) values at time/place *i*.

The result revealed that the model underestimated the LAI and yields both the cultivar and over estimated rest of the parameters. The average error percent of pod yield, anthesis and maturity for cv. Rajendra Mahsuri-1 as simulated by DSSAT model were 1.46, 1.94 and 1.62 respectively. The model successfully predicted growth, phenology and yield of crop with error values within 10%. Simulated growth and development (anthesis, physiological maturity, maximum leaf area index and grain yield) were close to observed and measured parameters for all the years of study. Based on these results it can be concluded that the model was very robust in predicting the critical phenological growth stages and yield of rice cultivar Rajendra Mahsuri- 1 in agro-climatic zone (IIIB) of Bihar (Fig.1).

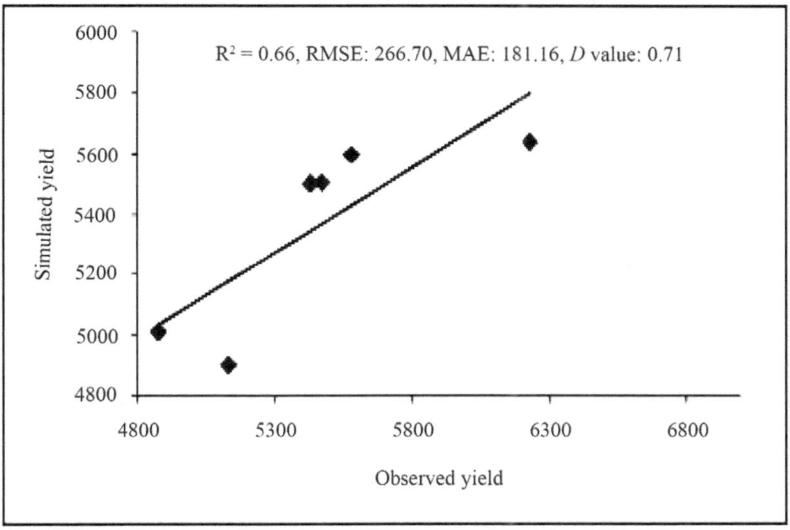

Fig. 1: Observed and simulated grain yield of rice crop at Patna Station

References

Aggarwal, PK and Kalra, N 1994. Analyzing the limitations set by climatic factors, genotype, water and nitrogen availability on productivity of wheat. II. Climatically potential yields and optimal management strategies. Field Crops Res, 38:93–103

Boote, K.J., Jones, J.W. and Burgeois, G. (1987). Validation of PNUTGRO a crop growth simulation model for peanut.proc.APRES, pp 19-40.

Directorate of Statistics and Evaluation. 2014. Planning and Development Department, Government of Bihar.

Timsina, J and Humphreys, E. 2006. Performance of CERES-Rice and CERES-Wheat models in rice-wheat systems. a review. Agricultural Systems 90:5-31.Effect of Foliar Application of Nutrients on Growth, Yield Attributes and Yield of Pre-Rabi Black gram

12

National Educational Policy 2020 and Agricultural Education An Overview and Way Forward

J.M.L. Gulati

Introduction

The much awaited new educational Policy 2020 approved and released by the Central Government on July, 29, 2020. A document which is set to transform Indian education system in most holistic way and will exert a long lasting influence on framing the Indian education system by addressing the vivid issues across the discipline and streams. The implementation of this policy will be under a single body HECI (Higher Education Council of India) well supported by multiple bodies like National Higher Education Regulatory Council(NHREC), Meta-accrediting body referred as Nation accreditation Council (NAC), Higher Education Grant Council (HEGC), General Education Council(GEC) also called Graduate attributes, it will formulate the National Higher Education Qualification Frame work (NHEQF). This council will take over presently regulating bodies of higher education like University Grant Commission, All India Council of Technical Education and National Council of Teachers Education. The professional councils like Indian Council of Agricultural Research, Veterinary Council of India, National Council of Teachers Education, Council of Architecture and National Council on Education in Vocational Training etc will act as Professional Standard Setting Bodies (PSSBs).

The NEP, 2020 intents to move forward in more holistic manner and multidisciplinary education mode(more visible at ground level). Even highly expertise and technical institutes will also have some vocational and value added courses, The policy envisages a three well defined and target oriented educational system like Research- Intensive Universities, Teaching- Intensive Universities and Autonomous Degree Granting Colleges. Besides, a well thought up shift from present 10+2+3 to 5+3+3+4 system of education.

Agricultural Production System

Spedding (1977) defined an Agricultural Production System as a system with an agricultural purpose and output. Haines (1982) considered agricultural system as food production system. The system is far more complex. It comprised of multidimensional components that interact in a complex way and influences the performance of crop. It operates with a purpose of achieving an out put through adoption of certain management tools involved in managing soil, growing of crops and act as fundamental production units.

Indian agriculture has gone through various challenges of diversified nature and recovered through temporary setbacks with outstanding strides that led not only revolutionized agriculture but sustained its growth and made the country a surplus state on food front from a deficit one. However, new challenges continue to pore in and few are knocking the door. The role of technology is much more relevant than ever before and the task is making farming activities more economically viable, knowledge intensive, more productive and sustainable. Though, the sustainability is always a question rather than an answer. Past experiences suggest that while disseminating the technology farmer to farmer learning was more acceptable to them. Blending of frontline technology with traditional wisdom will bring the technology more familiar to farmer, more appropriate, more acceptable and adoptable. It emphasizes that a technology should be affordable, accessible, and acceptable. Pinpointed alternatives and contextualizing the crop-research in integrated manner are needed for a holistic development and also to take care of long term ecological and human welfare and providing house hold food and nutritional security.

Agricultural Research and Education System

Agricultural Research

In Indian context, the agricultural research and education has come a long way in formulating research and educational policies over time and scale keeping in view the requirements and transformed Indian Agriculture and brought the country from deficit to surplus.

Some Milestones

1889- Establishment of Imperial Bacteriological Laboratory, Pune. Now Indian Veterinary Research Institute(IVRI) Izatnagar

1905- Creation of Agricultural Research Institute, Pusa, Bihar. Now Indian Agricultural Research Institute, (IARI)NewDelhi

1929- Establishment of Imperial Council of Agricultural Research. Now Indian Council of Agricultural Research(ICAR)

1966- Autonomous status awarded toICAR

1970- ICAR declared as Institute of National Importance as statutory body and conferred upon it certain functions performed by UGC in the field of agriculture. Assigned with the responsibility of professional guidance and regulation of quality of education.

1973- Set up of Department of Agricultural Research and Education(DARE)

1983- Making ICAR Tall, Slim and Healthy as recommended by a high power ICAR review committee with Dr. G.V.K. Rao asChairman

2004- ICAR to assigned to develop suitable models to actively pursue identified areas of research and other research related activities on commercial lines as suggested by committee chaired by Dr. A.R.Mashelkar

2004- revamping of National Agricultural Research System as per decision of Planning commission

Thus, over the years ICAR remained torch bearer of agricultural research in India. It formulated, executed, and transmitted the technological interventions to farming community through a vast net work of AICRPs, NWPs, commodity research and a large net work of KVKs.

Agricultural Education

Education is the backbone of a society- it can make it, it can destroy it. Policy makers on education thus, have a greater role in shaping the future of the nation. Agricultural education thus, is the backbone of agricultural production system. A trained, skilled and knowledgeable human resource is like a captain to sails his ship in right direction amid the odds.

Some Milestones

1949- Establishment of State Agricultural University. Recommended by 1st Education

Commission headed by non other than by Dr. S. Radhakrishnan

1952- Setting up of Education Panel

1962- Model act Of ICAR and coming up of 1stSAU

1963- Formation of standing committee

1966- Establishment of full fledged Division of Agricultural Education

1974- Establishment of Norm and Accreditation Committee

1996-Setting up of Accreditation Board

2007- National Agricultural Development Project

2008- New Model Act of ICAR

2012- Launch of National Agricultural Education Project

Agricultural education is key for achieving the daunting task of keeping the nation self sufficient which is possible through skilled, trained and knowledgeable agricultural graduates. The NEP 2020 has opened up windows overall and development of each wing- (teaching, research and extension) of agricultural production system. The agricultural education need to be comprehensive to match the demanding nature of NEP 2020.

Task ahead:

Let's take a leaf out from NEP 2020,

- The NEP 2020 envisage the adoption of 5+3+3+4, a shift from 10+2+3. And this change is important, important to think how to revamp agricultural education.
- NEP 2020 provides a three phase exit plan to the students. Issue of a certificate if one exits after 1 year, awarding diploma if exits after 2 years of study and a degree after completion of full term of 4 years. Another challenge to how to plan for this

Now the question is, is it applicable to a professional course like ours? The Indian Council of Agricultural Research declared UG degree in Agriculture and allied Sciences as professional degree vide F. No. Edu. / 13/1/2016-HRD dt. 06. 10.2016.

What next?

As the ICAR will be providing the professional guidance as per NEP 2020, the following few points are suggested:

- Constitution of a high power expert committee by ICAR to go through the NEP 2020 and suggest the role of ICAR.
- Formation of 6th Deans committee to suggest the relevant and appropriate changes in the course curriculum. It appears to be a daunting task and expected to have sea changes keeping in view if the exit plan becomes an integral part of the course programme.

- If so, the courses should be re modeled in such a way so that if student makes exit after one year of course, he or she should be able to justify the certificate. It required to provide skill and training and entrepreneurship-based and not so basics oriented
- If a student exits after 2nd year, they will be given a diploma. The courses should become more elaborative in nature and to raise the bar to bit higher level so that the diploma holders can match standard of present pass out from Agro-polytechnics.

Are our agricultural graduates really professionals?

As of now, I doubt that the students have understood it. I also reluctantly say that as a teacher, we lacked behind and more efforts are needed to make the agricultural students more competitive having expertise based knowledge and more importantly inculcate a sense respect to degree and the profession. The NEP 2020 provides an excellent opportunity to make this course more attractive, and a stream that one can choose for their future. Some suggestion maybe:

- Let's introduce agriculture course at 3rd phase of 9-12. Well under present system the students do read basic science courses. They read about plants. Let they read about crop. Let them know when a plant becomes a crop, its importance, and when one raises a crop, how to manage it. Means the students become aware of such an stream along with medical and engineering. The degree progrmmae in agriculture will then attract the students. They will join it by choice and not by chance.
- A three tire agricultural education system involving formulation of course curriculum, can be another option such as, Core courses of National prospective, Certain minor course having zonal relevance and Some supportive courses addressing the state level issues.

The policies towards agricultural research and education needs to be revisited to access the requirements making it more competitive and inclusive to address emerging issues and anticipated challenges.

Reorienting approach on capacity building

The NEP 2020, in itself, is demanding and challenging. The routine approach, just to produce the skilled and knowledgeable manpower will not suffice. The issue- based agricultural research approach has shifted from routine field and laboratory experiments to modeling and simulation mode. Use of sophisticated instruments has increased particularly in advanced countries, and in centers of excellence in our own countries. It is the era of smartness. Every now and then some new terms are coined suffixing smart; climate smart, energy smart, knowledge smart and so on. To cope up with this mindset we need to adopt student smart, researcher smart, faculty smart and extensionist smart approach. Accordingly, the support staff and the research staff at field level needs to be tuned to make them understand and craete enough capability to adopt such smart technologies. We also look forward towards smart farmers. Task is difficult and requires rethinking on agricultural education. The daunting food requirement of future under shrinking resources can only be met with capable manpower with new frame of mind. The course curricula should be in continuity mode up to the graduation level keeping focus on Post Graduate courses. The beginning has to be made from school level giving due importance to agriculture as subject. The students in general must know about this aspect of modern High-Tech Agriculture. The county would require about 333 million tonnes of food grain by 2050 for a projected population of 1.7 billion. It is not only the issue of food security but also of world peace as quoted by Dr. M. S. Swaminathan:

"Food has become a symbol of our collective human endeavor to create a better world for all. However, the victory has been partial, and neither the challenges nor the opportunities that food presents have been fully addressed. It is of crucial importance not only to the poor but also to the peace & stability of global society, that we complete the task of banishing famine & hunger once and for all. Hunger anywhere threatens peace everywhere".

Conclusion

Agricultural Education is the precursor of agricultural research and backbone of agricultural production system. Indian Council of Agricultural Research, New Delhi is the sole custodian to provide professional guidance to agricultural education across the country. With declaration of graduate degree in Agriculture as a professional one, there is an urgent need to change the approach and bring it on par with other professional course like medical sciences. The attitude of the teaching faculties needs to be tuned in that direction at 1st place especially in view of implementation of NEP2020. The education on basics of agricultural science should begin at school level (9-12) like any other science subjects. It will help the students to get acquainted with the stream at 1st place, and at second place, the course will attract bright students, those who will be really willing to take up this professional course. It will also provide more flexibility to planners of agricultural education in framing our semester courses, their breakup, their priority and the structure thereof. It will also help to arrange the courses keeping in view the three tier exit plan as envisaged in NEP2020. The courses needs tobe reoriented keeping in view the national, regional and states requirements in synergy with their respective research priorities

References

Annonymous. 2006. Evolution of Indian Agricultural Research System. In; Hand Book of Agriculture, Indian council of Agricultural Research, Directorate of Information nd Publication of Agriculture, ICAR, Krishi Anusandhan Bhavan, New Delhi: pp 1-16

Annonymous 2020. Transforming India : all about National Educational Policy 2020. Ministry of Human Resourse Development, Govt. of India, New Delhi

Speeding, CRW. 1977. Biological Efficiency in Agriculture. In: An Introduction to Agricultural System, Spriger. 1988.http://doi.org.10.1007/978-94-6408. Pp 52-72,Sundaram, V. Education in the new age.2003. Kisan World. 30(7) pp 6-8 .

13

Is Farmer Producer Organisations (FPO)-A Panacea for Farmers' Woes?

S. Shasani, G.S. Saha and H.K. De

Introduction

Agriculture as a primary sector in India is mostly production oriented. Global agriculture is experiencing significant changes in terms of technological and digital revolution, rapid urbanization, faster growth in middle income groups, changing food habits, market oscillations, and above all, climate change. During the last 50 years, India has recorded 4 to 10-fold increase in agriculture and food production(food grains-290 MT, horticultural produce -315 MT, milk production nearly 180 MT, and fish production approaching 12 MT), rendering the country as the second largest agrarian economy in the world (Singh, 2019). Despite the excellent progress, the country is home to over one third of the world's small and marginal farm holders, with scattered land holding. Eighty-five percent of the farmers are small and marginal. On the other hand, world's population is growing and to meet the growing food demand there is a need to increase agricultural output by at least 60 percent globally, and 100 percent in developing countries (FAO, 2014a).But, agricultural development does not depend solely on increasing production rather it should also focus on the social, economic and environmental issues. In this context, a sustainable solution can be achieved through collectivisation of agricultural produce and creating commodity-specific agri value chains with participation of agri- entrepreneurs and primary producers on the equitable terms.

Table 1: Advantages and challenges of Farmer Producer Organisations

Advantages	Challenges
• Economies of scale which reduces the cost of production • Better negotiation and bargaining positions • Vertical and horizontal coordination and value chain management • Enhance producers share in consumers rupee • Risk mitigation • Diminish cost of seeking information • Empowering small and marginal farmers economically and socially • Provides technical help in production and creation of social capital	• Huge equity capital requirements in starting FPCs • Lack of collateral security in availing loan • Huge tax burden • Weak value chain management infrastructure • Favoritism in organization • Lack of basic infrastructures • Lack of skill to maintain business records • Competition with rich middlemen • Lack of hand holding from state governments

(*Source*: Mukherjee et al., 2018)

What is an FPO?

A Farmer Producer Organisation (FPO) is a legal entity formed by primary producers, viz., farmers, milk producers, fishermen etc. Members are usually small farmers who are empowered by FPOs to trade in agricultural platforms. It is an organization of the producers, by the producers and for the producers. It is time to recognize the contributions of producer organizations and enhance their capacities to provide a much wider range of services to producers (Chander, 2013). Till 2018-2019, a total of 7,374 POs have been registered in the country. Out of these, 6391 producer companies are working on agriculture and allied activities, such as cultivation, plantations, dairies, non-timber forest produce, fish, poultry, etc. The top four states, Maharashtra, Uttar Pradesh, Tamil Nadu and Madhya Pradesh, account for 50% of producer companies registered in India (Govil et al., 2020). Government of India through Small Farmers' Agribusiness Consortium (SFAC) and National Bank for Agriculture and Rural Development (NABARD) is promoting FPOs by mobilizing the farmers and helping them in registering as companies.

In order to face these unique challenges, the present agricultural scenario must be transformed in to ecologically healthy, economically viable and socially inclusive one. Finance Minister, Nirmala Sitharaman in her budget speech (2020) announced that 10,000 new Farmer Producer Organisations (FPOs) would be formed in the next five years. Singh (2019) stressed on the growing agri-entrepreneuship to become a national movement to enhance rural income as well as employment security and helping to build a 5 trillion dollars economy. The number and types of such organizations supporting farmers with information, inputs and services have increased during the last three decades. Many of these actors are bringing varied skills and expertise that are needed for putting new knowledge into use (Sulaiman et al., 2010). Hence, there is ample scope for high value realization for produce through virtual market development by forming FPOs.

What is the need for FPO formation?

FPOs are increasingly considered a critical component of a rural economy's success, and many governments are beginning to recognize their benefits. Particularly in fragile states, rural organizations are crucial for bridging the gaps between public-sector provision of key productive and social services (Ragasa, 2014). They bridge the gap between production and marketing. FPOs play a major role in utilising the local knowledge and understanding to fit in to the needs of the communities. They are increasingly recognized as essential institutions for the advancement of agriculture, the rural economy and rural livelihoods. FPOs also strengthen the political power of farmers, ensuring the reach of their opinions to the policymakers and public. Economically, FPOs can help producers gain skills, gain access to inputs, form enterprises, process and market their produce more effectively to generate higher incomes (Hill, 2011).

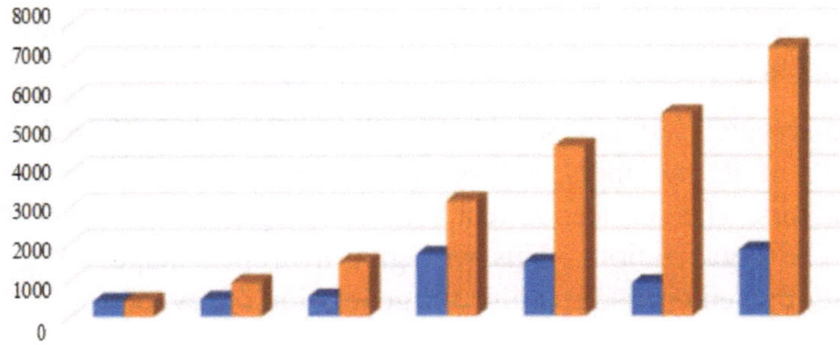

Fig. 1: Growth of Producer Organisations in India (Source: Govil et al.,2020)

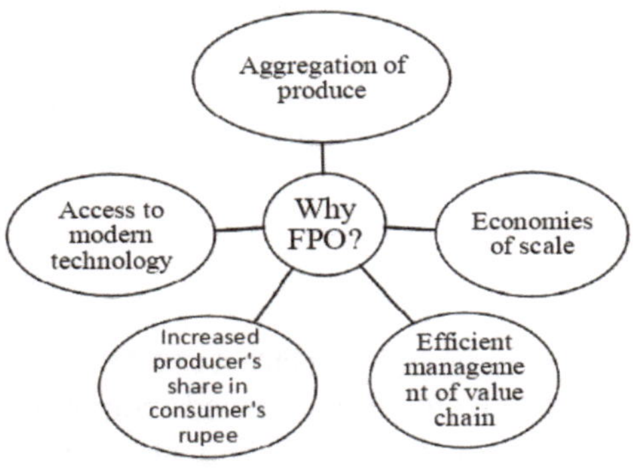

Fig. 2: The need for FPO

Key factors for a successful FPO (Source: FAO. 2014b)

i. Autonomy: Autonomy is the essential factor for long-term sustainability of the organization. Leadership, vision and initiative must come fom within the organization. Genuine decisions must be made rather than direct or indirect imposition by facilitators. Thus, the formation of POs cannot be an externally-imposed process, but must, instead, highlight the benefits of being organized, facilitate peer-to- peer learning from POs already established, and build on existing networks within villages.

ii. Inclusive leadership: The sustainability of the organization depends upon strong leaders that mobilize and engage their members effectively. Leaders must be genuinely endorsed by members and must make continuous efforts to engage and communicate with their organization. Grievance and accountability mechanisms should also be in place to address any concerns about poor leadership.

iii. A strong membership base: To build a strong foundation, members have to play a vital role. If the members are strong, its leaders are likely to be strong as well. Developing a strong, empowered membership base is critical for ensuring the sustainability of the organization. Members must be engaged in the affairs of the organization and must be able to hold their leaders accountable.

iv. Needs-based service: Services provided by the FPOs must be based on the true needs of its members, and not on the pre-conceived notions of external actors. The services reflect actual needs, ensuring financial viability of the organizations. This will give a proper shape for the development of an enterprise and also for generation of revenue.

v. A clear, 'owned' purpose: There should be a clear purpose for the proper functioning of the organization. This must give a precise message to all concerned farmers, and not just to the facilitators who, in many cases, impose their own ideas. This is a large factor contributing to the mechanical nature of PO development, and why organizations are at risk of failing upon closure of the project under which they were developed.

Conclusion

FPOs have come of age. With Government of India setting a target of establishing ten thousand more FPOs by 2024-25, the movement has gained momentum. The operational and managerial assistance provided by FPOs enable farmers to organise their production and post-harvest decisions. Besides farmers need to be educated about niche market and export oriented products. 'One district one product' need to be pursued in letter and spirit in order capitalize on the comparative advantage that the district offers. While developing policies for FPOs, it is important to consider the objectives on which the FPOs are established and also the target groups. The future strategies for scaling up of FPO should make an effort for developing and sustaining the institution. Such effort should be in the form of research, capacity building, policy and management support (Mukherjee et. al, 2018). FPO has huge possibility to become a model for enhancing livelihood security of small and marginal farmers in India. Sound business plan, innovative marketing, able leadership, need based service to members etc. will help FPOs tide over the challenges.

References

Chander, M. (2013, November 13) Producer Organizations (POs) and extension: The road ahead [Blog post]. (Retrieved from https://www.aesanetwork.org/producer-organizations-pos-and-extension the-road-ahead/

FAO. (2014a). State of Food and Agriculture. Innovations in Family Farming. Rome, FAO. (also available at: www.fao.org/3/a-i4040e.pdf)

FAO. (2014b). Farmers' Organizations in Bangladesh: A Mapping and Capacity Assessment. Rome, FAO. (also available at: http://www.fao.org/docrep/019/i3593e/i3593e.pdf).

Govil, R., Neti, A. and Rao, M. R. (2020). Farmer Producers' companies – past, present and future. Ajim Premji University, Bangalore, p 99.

Hill, C. (2011). Enabling rural women's economic empowerment: institutions, opportunities and participation? Ghana. Paper presented at the UN expert group meeting, organized by UN Women in cooperation with FAO, IFAD and WFP on Sep 20-23, 2011 in Accra,Ghana. (alsoavailable at: http://www.un.org/womenwatch/daw/csw/csw56/egm/Hill-BP-1-EGM-RWSep-2011_CH.pdf).

Mukherjee, A., P. Singh., M. Ray., Satyapriya and R. R. Burman, (2018). Enhancing farmers' income through farmer producers companies in India: Status and roadmap. Indian Journal of Agricultural Sciences 88 (8): 1151–61.

Ragasa, C. (2014). Improving Gender Responsiveness of Agricultural Extension. In A.R. Quisumbing, R. Meinzen-Dick, T.L. Raney, A. Croppenstedt, J.A. Behrman and A. Peterman, eds. Gender in Agriculture: Closing the Knowledge Gap. Rome, FAO.

Singh, R. B. (2019). Agricultural Transformation – A Roadmap to New India, National Academy of Agricultural Sciences, New Delhi, pp415.

Sulaiman, V.R., Hall, A., Vamsidhar Reddy, T.S. and Dorai, K. (2010). Studying Rural Innovation Management: A Framework and Early Findings from RIU in South Asia. RIU Discussion Paper Series #2010-11, Research Into Use (RIU).UK.

14

Agricultural Service Provider Vital Platform for Sustainable High-Tech Agriculture in Odisha

N.K. Dhal

Introduction

Agriculture sector is currently at the crossroads, which is evident in terms of higher nutritive food requirements juxtaposed with growing population. Food production faces multiple challenges due to frequent occurrence of natural calamities, sudden rise in temperature, increase in occurrence of disease and pests leading to both pre and postharvest crop loss. Alarming decrease in fertility status of high and medium land soils of Odisha due to formation of hard clay pan below 4-6" of soil depth has resulted in slowing down of nutrient mineralization by the soil microorganisms within the soil. Crop production stress is further aggravated due to steady decrease in availability of cultivable land with the advent of urbanization, which concurrently reduced the number of dedicated agricultural households. It has been estimated that production of all food grains and horticultural crops needs to be doubled at farmer's level in order to feed the growing population of the world failing which there will be food crisis by2050.

Socio-economic Status

Doubling of food production at farmer's level is not at all an easy task as many factors relating to climate change, crop stress conditions (as mentioned above) and socio-economic condition of farmers are adversely affecting agricultural production. Unlike their forefathers, new generation youths of farm families are less inclined towards farm activities. The farmers living in the urban fringe are mostly considering agriculture as their secondary occupation and evince interest in retail businesses like shops, services in various government and private organizations, contract services, social works etc. in order to meet their family's financial and social needs. A rising trend is observed in land leasing to the informal tenant farmers by the medium and large landowning farmers or they prefer to keep their land fallow to avoid litigation. The farmers cultivating the land on leased basis are reluctant to adopt high-tech agricultural practices. The above observations are based on a study conducted by the author while working in one of the adopted village, Champati of Padmapur block for three years. Another observation is that land owning affluent farmers in their quest and curiosity for profitability resort to high tech agricultural practices and are investing large sums in commercial agriculture like coffee, tea, fruit orchards etc. The need of the time is to double the production of food grains such as, cereals, pulses, oilseed; vegetables and fruits which at the same time must be highly profitable.

Public sector Initiatives in Providing Agri-services

Government at different times has introduced schemes such as, Rashtriya Krishi Vikas Yojana (RKVY), Pradhan Mantri Krishi Sinchai Yojana (PMKSY), National Mission on Oilseed and Oil Palm (NMOOP), National Mission on Agricultural Extension and Technology (NMAET), National Mission for Sustainable Agriculture (NMSA), Rural Infrastructural Development Fund (RIDF), Chief Minister Samagra Gram Unnayana Yojana (CMSGUY), Agriculture Technology Management Agency (ATMA) for benefiting small and marginal farmers through advisory services provided under such government schemes. Irrespective of the outreach of the scheme, the agricultural growth is still hovers around 2.0 percent and is not enough to feed the total population of the world by 2050. Research and extension efforts to motivate farmers for precision farming and/or technological innovations have become unworkable, except for few adoptions. Similarly, farm machineries supplied under the Sub-mission on Agricultural Mechanization (SMAM) scheme to the farmers with a subsidized price are not being used exclusively for agricultural purposes. The foremost example is tractors, which are mostly used for transport of materials.

Profitable Agriculture

Agriculture as a profit making venture depends on many factors such as: proper crop planning for adoption of intensive agricultural practices; measures to improve soil fertility through proper land preparation; application of Farm Yard Manure (FYM) and green manures; timely availability of inputs; balanced and timely application of fertilizer; timely irrigation; judicious plant protection measures; technology based cultivation measures; measures to reduce post harvest losses and forward linkage for sale of the produce or value addition to maximize profit. All the aforementioned measures are vital for sustaining production and the key to higher production is timely application of technology and crop planning. However, we confront many technological gaps in our production chain, which necessitates right application of technology and planning in agriculture production at right time. In this regard, making technology accessible to the end users, to help them carrying out agricultural operations starting from land preparation to harvest using an integrated approach is the critical need for making profit sustainably. This can be achieved through institutionalizing a cadre of Agro-Service Providers at Gram Panchayat level which should be able to provide technology based agricultural services on contract basis.

Present Status of Contract Farming

Presently, contract services in construction, security, sanitation, garden maintenance are in vogue and are widely promoted by both private players and government agencies. It is providing service opportunity to unemployed persons and renders better quality of service. Individuals or organizations hire different service's depending on their financial strength. Now it is imperative to encourage entrepreneurship in agricultural services and give agriculture the status of an 'industry' without much financial liability to the government.

Agro-service provider could be part of an agency (privately owned and managed) which provides various agricultural services beginning with crop planning to processing and marketing. The concept of Agro-service provider is not new. There are various government, semi government and private agro-service providers working at village level which provide technical information and support in various aspects with regard to farm machinery, seeds, plant protection, agricultural finance, soil testing etc. separately or in combination of few aspects together. The existing government extension system cannot solve the problems of individual farmers at one time, which enhances the scope for accessing services from various agro-service providers for supply of inputs like, seed/ seedling, fertilizer and plant protection chemicals suppliers to overcome their problems. Many times it has been seen that the problems get more aggravated due to excess application of such imputes resulting increase in chemical toxicity and decrease in nutritive value of food produce. In some agriculturally advanced states like, Tamilnadu, Maharashtra, Gujarat, Andhra Pradesh, Karnataka etc. the agro-service providers are offering valuable services and equate agriculture to industry with sound technical manpower. As a consequence of COVID-19 outbreak, unemployment in rural farm sector has emerged as a great challenge before the state governments all over India which can be partly be solved by uplifting agriculture to the level of industry.

Agro-service Provider (AGSP) – A Single Window System

Often, it has been observed that simple ideas turn into success stories with far reaching impact. In Odisha, it is time to establish Agro-Service Provider platforms in each Gram Panchayat on a PPP mode, with the mandate to carry out total operation for a crop on leased land or any farm work starting from summer ploughing to harvesting of a crop on hiring or on contract basis with reasonable price tag for each operation. Such a system will ensure timely operation of agricultural practices for each crop and create pathways to develop as a model centre for high-tech agriculture. Also, it will create employment opportunities for educated village youths and farm women at panchayat level, while ensuring timely agricultural operations, the key to higher production. The state government has shown deep concern for development of agriculture through introduction of new agricultural policies from time to time. The proposed Agro-service Provider platform may be considered for inclusion in the new agricultural policy for the benefit of large to small land holders who partly depend on alternate sources of income for their livelihood support.

Need of Provisions to be Specified in the Agro-service Providers (AGSP)

1. Each agricultural operation for a particular crop must have a fixed time limit. Example: for rice, summer ploughing should be completed between 15thApril to 31st May and so on so forth.

2. Each AGSP must have at least two hectors of land in their possession with infrastructure facilities for farm machinery unit, agricultural input sale center, nursery unit, high-tech precision farming unit and half of the land should be utilized for cultivation of multiple crop using crop rotation practices suitable for thelocation.

3. Each AGSP unit should have provision for all the farm machineries and man power support to carry out any farm activities beginning with land preparation to harvesting and processing/value addition of the farm produce. Such services must be provided either on hire basis or oncontract.

4. The AGSP may develop an operational business model and fix price for each services. The service costs may be displayed along with economics of crop production in order to explain the benefit of the system. Online platform should be created and made accessible for individual farmers for registration of date, time of operation and payment option for each farm activity.

5. The AGSP should have the option to cultivate the farm lands, if available, on lease basis at least for one year starting from 1st June to 31st May of next year with a crop planning to raise at least two crops. In such case, the AGSP must guaranty the benefits to be accrued to land owners which shall be calculated on the basis of size of the land, type of crops to be grown and location of the land.
6. The AGSPs could make financial provision to buy back the produce from the farmers and supply it to the market/ value addition chain.
7. It may have the training facility for the farmers apart from resources for organizing field days to demonstrate technology and practices to a large number of farmers.
8. The ASGPs should be managed by an Agri-business Management personnel or a graduate having completed a short course motivational training on Agri- business.

Conclusion

The policy and implementation challenges confronting sustainable growth in agriculture are evident with government amplifying its actions towards doubling income from farming, retaining youths (particularly next generation farmers) in agriculture and addressing the gaps in enforcing timely agricultural operations, input supply and agricultural mechanization. The Agro- service providers offer a platform that will potentially address the aforementioned concerns and a self-replicable model which will contribute to better adaptation of technology and farming models. The AGSPs can play the role of facilitator and executor in demonstration, training and incubation of technology and practices and will contribute to production, value addition and marketing efforts thereby enhancing the economic returns from agriculture. When it comes to financing of production and processing activities, the AGSPs as aggregators will actas guarantors for the financial institutions. Thus, AGSPs would facilitate transfer of knowledge, technology and financing, considered most vital instruments for sustaining agriculture. Considering the recent policy announcement “One Nation One Market” for agricultural commodities, the return from agriculture may grow only when such AGSPs operate as catalyst for change and support farmers/farming community in containing the input cost on farming and help the farmers in realizing better profits from the output.

Acknowledgement

Deeply acknowledge the valuable input given by Mr. J P Mohanty, The World Bank, New Delhi and Prof. DP Ray, Former VC, OUAT. For their valuable guidance in designing the AGSP.

15

Assessment of Watershed Projects in Tribal Districts of Odisha

S.K. Behera

Introduction

Agriculture is the main sector of the Odisha economy and contributes approximately 42% to the gross domestic product (GDP) and employs over 80% of the population. Despite its role, agricultural production is constrained by high climate variability where rainfall distribution is extremely uneven and this has negative implications for the livelihoods of people .Drought frequently results in crop failure, while high rainfall intensities result in low infiltration and high runoff causing enhanced soil erosion and land degradation. Land degradation in the form of soil erosion and declining land fertility is a serious challenge to agricultural productivity and economic growth (Lemenih, 2004).Studies indicate that the tribal district of Odisha have experienced high rates of soil erosion and deforestation, resulting in sediment accumulation in downstream reservoirs and rivers.

In general, watershed degradation resulted in long-term reduction in the quantity and quality of water and land resources, which negatively impact on the livelihoods of the rural poor This spurred the Odisha government to launch an extensive soil and water conservation (SWC) program, which began in the early 1970s. In response to the frequent drought in the northern and western part of the State during the period in every four years the World Food Programme (WFP) supported the Food for Work (FFW) project, which was launched in 1974 initially as an emergency relief initiative later evolved into a development. Participatory integrated watershed management approach emphasizes improving the productivity of water and land resources in an ecologically and institutionally sustainable way (Farrington et al. 1999). Hence, watershed management has become a central point of the rural development and poverty alleviation agenda. According to the participatory watershed management guidelines (L. Desta et al. 2005), the objective of watershed management is to improve the livelihoods of rural communities and households through

Productive uses of SWC

- Rainwater harvesting for improved groundwater recharge.
- Promoting sustainable farming systems and agricultural productivity adopting suitable soil, water, nutrient and crop management practices.
- Rehabilitating and reclaiming marginal lands through appropriate conservation measures, such as planting of trees, shrubs and grasses depending on existing potential.
- Enhancing the income of smallholders by diversifying agricultural practices.
- Income -generating activities (IGAs).

Project Area

The State of Odisha endowed with rich natural resources, is also one of the poorest states. The Schedule Tribes (ST) and Schedule Castes (SC) constitute nearly 40% of the population of Odisha. There are 62 Tribal groups and 13 particularly vulnerable Tribal groups (PVTGs) living in the state, a majority of whom are economically, politically and socially backward. In spite of several interventions by both the State and Central Governments, the Tribal communities are far behind the so called mainstream of development. Focusing on the STs, the Government of India and State Government jointly sought assistance of IFAD for its financial support for the holistic development of the Tribal communities in the state. The Project was implemented in 30 backwards blocks of seven districts Kalahandi, Kandhamal, Koraput, Gajapati, Nabarangpur, Malkanagiri and Rayagada of Odisha in 358 micro watersheds covering and treating an area of 175,358 hectors .The target households include 75% of ST,14% of SC and 11% of other communities along with 70% BPL house hold.

Odisha Tribal Empowerment and Livelihood Project was implemented in 30 Blocks of seven districts by the respective Integrated Tribal Development Agencies (ITDAs) in the south-western region of Odisha covering 56180 households in 358 micro-watersheds and 1 042 villages. These Blocks have high concentration of tribal households. Poverty rate of these Tribal communities at baseline was 63% .The project area is in remote locations devoid of critical infrastructure such as road network, water supply, markets, credit, etc.. Major parts of the programme areas in the southern tribal belt are plateaus at altitudes ranging between 800 and 1000 m. Large variations in hydrology and vegetation within short distances create a mosaic of extremely diverse environment. A characteristic climatic feature of the area is that its proneness to drought, not so much due to a scarcity of rainfall (the area receives 1040-1400 mm of rainfall per annum) but its extreme unreliability. Dry-spell of 15-25 days can destroy crops during the main cropping season of June to October. Nearly 76% of farmers are small farmers cultivating less than 2 ha of land and over 50% of farmers have less than 1 ha of land. Landlessness is high (20-30% households with absolute landlessness; over 50% landless households as per definition of the state government) but a sizable number of them are engaged in share cropping. Agricultural productivity is low with an average productivity of 350 to 450 kg/ha of rice, one of the major crops of the project area.

Access to land is the key defining factor in vulnerability to poverty. On average, agriculture provided households with sufficient food for a maximum of six months in any normal year whilst those households that relied on shifting cultivation (podu) their production covered barely one to two months. The reason for seasonal food shortage was mainly due to low productivity of land. Ecological imbalance has been seriously undermining the livelihood patterns of tribal households and increasing their vulnerability. Shifting cultivation has also been leading to severe soil erosion, loss of soil fertility and declining productivity of crops. With their increased dependence on land and agriculture, lack of title to land and the related potential for dispute was a major source of tension in the tribal area.

Methodology

The Study adopted a quasi-experimental design along with a mixed approach of combining quantitative and qualitative methods viz. household survey, focused group discussions interviews with key stakeholders and assessment of physical interventions .Multistage random sampling was adopted for selection 2017 households and the data collected from the respondents were by using CAPI (Computer assisted Personal Interviewing)tools. The members of the experts team visited selected villages and had interactions with individual beneficiaries, groups ,projects-supported village-institutions, the staff of the FNGOs and the ITDA and other government departments. Findings are presented in terms of achievements against agreed projects (output, outcome) and standard development indicators and on analysis of project management and normative concerns. Establishment of theory of change by comparing with baseline and control and causality attribution have also been made.

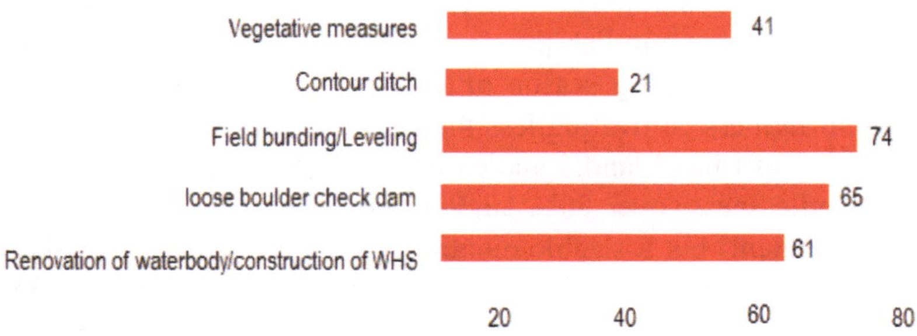

Fig. 1: Participation in land & water Management Measures

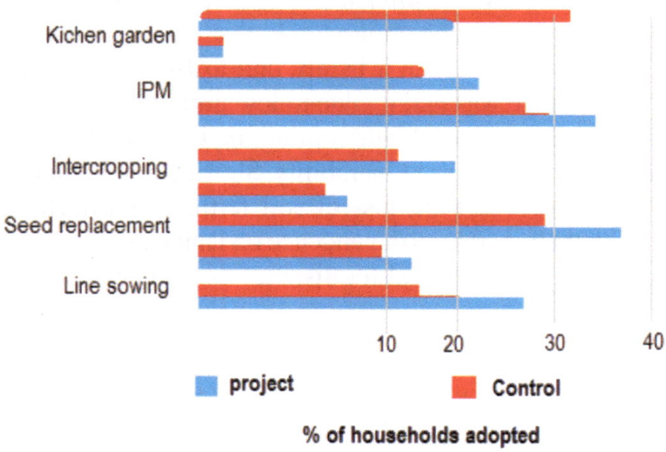

Fig. 2: Adoption of improved agricultural Practices

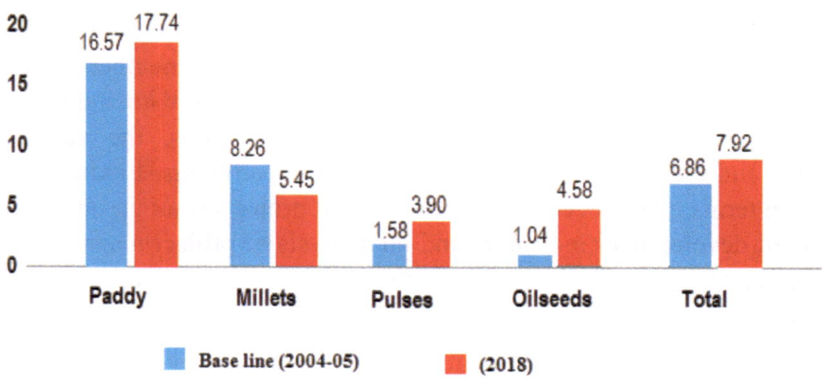

Fig. 3: Average crop Yield per ha in quintal

Result and Discussion

The key findings of the impact assessment in different sectors like land and water management, agriculture practices, productivity, yield and income revealed that:

- About 175,368 ha of land in 358 micro watersheds areas were treated.
- 86% of households were directly benefitted.
- There has been 22% increase in arable land and 8% increase in irrigation area.
- ORSAC analysis indicated about 34% reduction in fallow land.
- 54% rate of the structures were found satisfactory.

In case of adoption of improved agricultural practices,

- This intervention has benefitted over 81% of the targeted households.
- 77% of the farmers are using such technology.
- Double cropping area has increased by 54%/ ha as per ORSAC analysis.
- Farm income has increased from Rs4756/ in the year 2005 to Rs.9785/ ha during 2017.
- Farmer's productivity has increased from 0.69t/ha to 0.79t/ha.
- 37% of households attribute yield enhancement to adaptation of practices recommended
- Cropping intensity has increased by 2.7% against 5.4% decrease in the State for the same period.
- Crop diversification index is 0.67which is higher than the state average of 0.52.
- Double cropping has increased by 50% with summer crop area increasing to almost three times.
- There is enhanced soil fertility and organic carbon stocks in the project area than the control.
- Quality of life, a composite index of access to clean and safe drinking water, access to clean fuel, use of toilet, electricity connection and malnutrition, shows better quality of life in the project area marginally over the control areas

Conclusion

For successful watershed management and effective up-scaling of best practices, this study recommends the following:

- Develop technologies that are suitable for a specific watershed and small holder farmers. Emphasis should be given to technologies that improve conservation of the resource base, while eliciting short-term benefits for the local people
- Improve surface and subsurface water management to increase water productivity
- Develop institutions and mechanisms that improve partnerships among stakeholders
- Devise mechanisms that ensure fair and equitable division of costs and benefits, especially between upstream and downstream communities
- Ensure community participation with adequate technical and financial support
- Link physical and biological conservation activities with IGAs and livelihood improvement activities
- Improve access to markets as high-value commodities are produced under watershed management having limited access to markets and post-harvest management technologies. The majority of the high-value commodities are usually perishable having sort post harvest life. Therefore, there is a need for effective integration of production, marketing and agro-processing to obtain the maximum benefits from the value chain.

References

Annual report of OTELP,20014,2015,20161nd,2017, Govt. of Odisha

Asefa, D.T., Oba, G., Weladji, R.B.,Colman and J.E. 2003. An assessment of restoration of biodiversity in degraded high mountain grazing lands in northern Ethiopia. Land Degradation & Development 14(1): 25-38.

ATA (Ethiopian Agricultural Transformation Agency). 2013. Annual report 2013/2014: Transforming agriculture in Ethiopia. Addis Ababa, Ethiopia: Ethiopian Agricultural Transformation Agency (ATA)..

Desta, G., Nyssen, J., Poesen, J.Deckers, J.,Mitiku, H.,Govers, G. and Moeyersons, J. 2005. Effectiveness of stone bunds in controlling soil erosion on cropland in the Tigray Highlands, northern Ethiopia. Soil Use and Management 21(3): 287-297.

Desta, L.; Carucci, V.; Wendem- Ageňehu, and A.Abebe, Y. (Eds). 2005. Community based participatory watershed development: A guideline. Addis Ababa, Ethiopia: Ministry of Agriculture and Rural Development (MoARD).

Farrington, J, Turton, C, James and A.J 1999, Participatory Watershed Development: Challenges for Twenty first Century., Oxford University Press, New Delhi, India.

Murty, V, V, N and. Jha, M.K.2013, Land and Water Management Engineering, Sixth edition, Kalyani Publishers, New Delhi.

Ramakrishna, P.S, 2004, Ecology and Sustainable Development, 2nd edition, National Book Trust of India

16

Gender Mainstreaming: A Strategic Approach for Achieving Gender Equality and Empowerment of Women

L. Das

The term 'gender' was first used by Ann Oakley and others in 1970s to describe the characteristics of men and women which are socially determined in contrast to biological differences. The term 'gender' further explains, a neutral term meaning, either men or women or both are in a particular context. The goal is to improve the status of disadvantaged class and get rid of socially created and approved discriminations. The green revolution in India changed the face of agriculture but apparently it contributed to two general trends (FAO, 1996) where wealthy have benefited more from technological change in agriculture than the less well off and secondly men have benefited more than women. In general, it is true that the priorities are set by men and some women, who are unaware of gender issues, and that these are seldom considered unless the research is sociological. But even social research can be totally gender blind.

Agriculture is the primary source of income for women in India which contributes more than 80 per cent of their livelihood of which 33 per cent belongs to agricultural labour force and rest 48 per cent are self-employed farmers. Globally 43 per cent of women are involved in agricultural activities; they are referred as the backbone of agriculture. Women are the major producers of food but they lag behind men in ownership of land and access to income from land (FAO, 2011). More than 70 per cent of rural women workers are engaged in agricultural work, (NSSO, 2017-18). In contrast to their high participation, women own 13.9 per cent of landholdings (Agricultural census, 2015-16). Globally, women receive only 5 per cent Extension Services (FAO, 2012), only 15 per cent of the world's extension agents are women and 10 per cent of total aid for agriculture, forestry and fishing goes to women. If we compare women workers in agriculture then we can see women workers are more involved in non agricultural activities (65%) than agricultural activities (35%). There is a decrease in number of cultivators but increase in number of labourers over the decade.

Moving onto Gender Gap, As per World Economic Forum, India despite of achieving high rates of economic growth in recent year was ranked at 140th position among 156 countries in Global Gender Gap Index in 2021 as against 112th rank in 2020. The gender gap in India has widened to 62.5%, largely due to women's inadequate representation in politics, technical and leadership roles, decrease in women's labour force participation rate, poor healthcare, lagging female to male literacy ratio, income inequality.

The report is a measure of gender gap on four parameters: economic participation and opportunity, educational attainment, health and survival, and political empowerment. The index has benchmarked 156 nations across the globe in 2021. The data show that it will take 135.6 years to bridge the gender gap worldwide and the pandemic has impacted women more severely than men. The gap represents an 'extremely high' level of gender inequality thereby resulting in its poor performance on gender related assessments despite being a large geographical entity with enormous natural resources.

Gender based discrimination continues in multiple ways: women are not recognized as farmers in Indian policies thereby denying them of institutional supports of the bank, insurance, cooperatives, and government departments. Unequal landholdings, social barriers relying on traditional methods and lack of extension facilities are some other challenges faced by women. Women plights are never ending in agricultural sector exploitation by middlemen, low wage given to small labourers, heavy discrimination and violence by male counterparts and the impact of technology have suppressed their voices. The ones whose day start before sunrise and continue after sunset, their voices are trapped in four walls and often go unheard owing to their gender. Their voices need to be heard to realize the dream of a progressive India.

Women contribute a significant share in agriculture but their growth and development have been shackled by multiple challenges faced by them. Increasing gender stereotypes, male-centric technology development, decreasing water availability, natural resource degradation, increased small holdings, climate related challenges, have become for farm women in agriculture need to be addressed and increasing population have affected tremendously. India, like many other countries of the world, has the onerous responsibility of achieving the Sustainable Development Goals (SDGs) by 2030. Out of the 17 SDGs set out by the United Nations, where Goal 5 states Gender equality & Goal 10 states reducing Inequalities. The gap can be closed and the issues can be resolved through appropriate policies and government programmes. Priorities need to be identified and strengthen the linkages that will boost socio-economic status of women.

Gender Gap can easily be identified through certain constraints such as lesser access to productive resources as compared to men, granting fewer and smaller loans, only 12.8 per cent of land holdings despite women being crucial in agricultural process. Assessing technology in agriculture, 5 per cent out of 45 per cent agricultural workforce have access to agriculture extension services. If women had equal access as men in developing countries then the outcome would have been different; yields in individual farms can increase by 20-30 per cent, overall 4 per cent agricultural production will increase and ultimately the number of undernourished people will be decreased by 12-17 per cent.

Gender Mainstreaming

At First we need to understand what is gender mainstreaming? It means integrating a gender equality perspective at all stages and levels of policies, programmes and projects. Women and men have different needs and living conditions and circumstances, including unequal access to and control over power, resources, human rights and institutions, including the justice system. Secondly, why we need it? It is a strategy to improve the quality of public policies, programmes and projects, ensuring a more efficient allocation of resources. Better results mean increased well-being for both women and men, and the creation of a more socially just and sustainable society. Finally, how gender equality issues need to be mainstreamed? Gender analysis and gender impact assessments are crucial tools for gender mainstreaming. These tools support the practical implementation of gender mainstreaming. Other factors are equally important to ensure proper gender mainstreaming, such as political will, commitment to and awareness of gender equality issues, knowledge, resources (including expertise) and availability of information.

Gender Budgeting has become a powerful tool for gender mainstreaming. It is used as a tool for effective policy implementation where one can check if the allocations are in line with policy commitments and are having the desired impact. Why focusing on gender? Because women represent 48 per cent of total population in country, women face high disparities in access to and control over services and resources, women's need to be more fulfilled in more strategic ways (gender blindness should be eliminated) and practical and strategic gender needs to be addressed.

Involvement of Women to Impact on Agriculture and Livelihood

Improved farm women's access to agri-production resources would enhance women's efficiency and increase agricultural productivity. Frequent involvement will result in increase in adoption of improved farm technologies and improve household food and nutrition security. If women had provided the same access to productive resources as men, there will be boost in yield by 20-30%, rise in agricultural output by 2.5-4% and subsequent reduction in hunger by 12-17% (FAO, 2011), When women and men are relatively equal, economies tend to grow faster, Poverty is reduced significantly and enhancement of wellbeing of men and women (World Bank, 2012).

Gender Equality, Poverty Reduction & Inclusive growth

The challenge faced by policy makers in addressing gender gaps and systemic constraints is the complexity and inter-connectedness of the wide range of policy interventions and multiple paths to reform, since "gender-differentiated market failures, institutional constraints, and persistent social norms often combine to reinforce gender inequalities and make improving gender equality much more complex. When there are multiple constraints, they all need to be addressed." (World Bank, 2011). Proper policy formulation and implementation would result in gender equality, and increase growth and resilience.

Strategic alliances creating Research-Development-Policy Interface

Certain factors are required to create a Research-Development-Policy interface such as initiating capacity development programmes, joint learning, application and practice of recently learned techniques, enabling environment for wider application and effectiveness of the learned techniques. Apart from these an interface for women needs to be formulated *viz*., P, Q, R, S, & T where P depicts production, productivity and profitability; Q for quality and quantity, R indicates remunerative and research, S reflects sustainability and supportive and lastly T suggests training, trading and technology.

Concept of Gender Sensitive Entrepreneurship Model through Institute-Industry-Stakeholders Linkage in Convergence Mode

A refined and women centric entrepreneurship model was developed for farm women at ICAR-CIWA where they boost their self confidence, update their knowledge and improve their skill as an innovator, a learner and a leader. In stage II, ICAR-CIWA, NGOs, successful women groups, WFPOs create a gender friendly environment to improve self actualization by initiating 4C's approach. In stage III, ICAR-CIWA, KVKs, SAUs, NGOs and state line departments engage in capacity development and skill development programmes for farm women. In stage IV, ICAR-CIWA, state line departments and industries educate farm women on recent ongoing farm technologies. In stage-V, where policy intervention, such as, government subsidies and schemes come into force. State line departments, NABARD, NHB, and APICOL etc. provide assistance. NABARD, Banks and cooperatives provide finance of any kind of entrepreneurial activities carried out by farm women. Finally the market linkage plays significant role in completing the cycle where farm women as an entrepreneur sell their produce/ value added produce directly without any middlemen intervention. Market linkages are often linked to big industries, for e.g., Agro-based food processing industries.

Recommendations and Conclusion

There is need to generate educated, trained, self-reliant, self-motivated, innovative and visionary women farmers to lead our agricultural production systems out of their multiple responsibilities. For '*Atmanirbhar Bharat*' it's important to take women along. The focus should be on the role of women in reviving the economy and making India self-reliant. Hence, a gender transformative approach is required to transforms gender inequalities addressing their root causes. For e.g., enhancing women's access to and control over resources. A gender accommodative approach/reforms has accepted existence of gender inequalities without addressing their root causes. eg., providing tubular maize Sheller to reduce women's drudgery in maize shelling is also required.

Gender Responsive Strategies

i. Make development institutions gender sensitive- Incorporating gender dimension in development approach, making an explicit mention of gender in mandates and policy, recognizing women as a stakeholder, providing opportunities to encourage women's participation.

ii. Enabling development managers and planners to understand and apply gender perspective in respective areas such as Gender sensitization, Gender responsive budgeting and Gender Equity

Gender Integration: The 3Es Concept

Reach

a) Include Women Growers, Labour, Girls and Entrepreneurs in program activities.

b) Number or proportion of women participating in a project activity.

Benefit

a) Increase women's well-being (e.g., food security, income, health) and skill.

b) Positive and negative outcome indicators such as productivity, income, assets, nutrition, time use etc.

Empower

a) Strengthen ability of women to make strategic life choices and to put those choices into action.

b) Women's decision making power, reduction of outcomes associated with disempowerment.

Hence, a 3E's concept was developed for bringing gender equity, promoting entrepreneurship and realization of empowerment with planning for training, research and extension activities for farm women development and federating farm women into business group. Concentrate on skill development, articulation of Tradition, Technology, Talent and Trade and developing farm women based clear road map for agricultural development in the country are required for bringing women to mainstream in agriculture and allied sectors.

17

Perspectives of Dedicated TV Channel for Farmers of Agriculture and Allied Sector on 24 X 7 Mode

B. Parasar and N.K. Dhal

Introduction

In the era of communication and information, each one of us should have access to information as per our choice, interest, priority, need and requirement. We all are playing different role in different walk of life in different capacity, for different professions, avocations, duties and responsibilities as well. To internalize any concept or idea, a perceptual clarity is very much necessary for interweaving it in our experiential textures. This is possible when more and more number of sensory organs are engaged in a learning situation irrespective of the subject or content to learn. In 21 Century with the advancement of science and technology, more particularly, Information and Communication Technology (ICT), it is possible to establish the connectivity on e-mode and do the things on virtual mode to the possible extent. It is quite possible in urban set up with proper infrastructural facilities both at the source as well as at the receiver end, without any distortion of message on account of electricity failure or wrongful handling of computer or any gadget for that matter. But this is not the situation when we think about unapproachable, inaccessible, unfrequented areas of hilly terrain or desert or adjacent areas of dense forest-a kind of complex, diverse and risk-prone areas at the country site. Thus, many people are sometime deprived of many valuable information in the context of their decisions, relating to the their professional, personal, vocational and social life.

If we look into the country's economy which is seriously banking upon the growth and development of agriculture, the very primary sector of Indian economy, over which the other two sectors, Industry-the secondary sector and service-the tertiary sector rest, has to be handle with care, precaution and very dedicatedly for the purpose. As to the extent of 70% of work workforce are engaged in agriculture, and in the process other stake holders also get involved in the game of production and productivity of agriculture and allied sector. The basic factors of production in addition to inputs like seeds, fertilizers, pesticides and money too demand most important input that is information for the farmers for the sustainability of farming enterprises. In common parlance it is said, if we read, see, write and experience the things, then we will know it, understand it. And if we know it, then we will find ourselves in win-win situation. Thus a person who is well informed then he or she will ahead of others but If the person is ill informed or poorly informed then he or she will be lagging behind others. Then how to make him known about the subject, content, idea, concept, and information?

Challenge Ahead

The knowledge is also viewed as retained information concerning fact, concept and relationship, Again knowledge is also further divided as awareness knowledge, How-to-Do type of knowledge and principle knowledge concerning any innovation or any such newly developed technologies or solutions pertaining to the problems that the people confronted with. So, it is a big challenge before all of us those are involved in the affairs of agriculture. On one hand we are flooded with information at level of the source and at the other hand there is a serious knowledge gap or communication gap between source and receiver; at the level of end users of such knowledge, and they may be our very respected, very lovable, very esteemed farming community if we view it in the context of domain of agriculture.

Story of Television in India

We look back to the history of mass media for a heterogeneous, anonymous mass audience, then, generally we may agree that 50s, 60s, and 70s, we all witnessed the spectacular performances and significant contributions electronic mass media in communication revolution. Gradually television came to the picture more prominently during early 80s, 90s and in the year 1982 during ASIAD GAME organized in our country and slowly the medium became popular with the telecasting of cricket match in our country more particularly during the time of World cup. Then the habit of viewing TV developed initially with the urban viewers and later penetrated gradually to rural areas and captured the rural mass audience with various of programmes like News, Cinema, Dance, Songs, Drama, Serials, Scientific discourse, food, Religious Discourse and also Educational Programme on behalf of UGC in addition to covering and highlighting places of religious interest and tourist place, the biodiversity, the flora and fauna, and bit of our rich cultural heritage blended with or supported by commercial advertisement during telecast of either live programme or recorded programme as the case may be through exclusive channels.

Towards the end of 20th century those television channel too incorporated agriculture information based programme both by Door Darshan, E-TV, and many other private channels, At the beginning of the 21" century, also those who were in the helm of affairs of this electronic media, felt the impact, importance and effect of agricultural programme upon farming community and started concentrating / focusing more on this type of programmes. But, soon after the habit was developed on the part of the television viewers and more particularly with the farmers showed interest to view such programmes. The programme relating to agriculture is of very short span, which may not always coincide with the conveniences of the rural viewers with regard to their available time and physical presence in the place where televisions sets are kept for viewing the programme when such programmes are being telecasted.

Importance of Aoricuatural Information and Technology

In Farmer's Lives

Just like we need for all the time air to breath and water to drink at any point of time when feel thirsty for our survival and sustenance, agricultural information, relating to any important tools and techniques methods, procedure, success and any other facts of agricultural sector should be available to the farmers irrespective of any particular time slot or any point of time in 24 hrs of a day. They can access it as per their conveniences' irrespective of a particular day or during particular time slot of the day.

It is also pertinent to know and understand the finer details as well as the intricacies exist in technology transfer process for appreciating the role of TV channels. Technology has been defined by Dr. V. E. Subaratnam in (1987) as the research findings which are having some practical utility, tested and found economically feasible, socially acceptable, culturally adaptable, scale neutral, gender neutral and eco-friendly as per the farmers condition and also legally tenable.

Technology should be understood not only as a process but as an amalgamation of skills, methods, techniques and equipments. On the other hand, appropriate technology refers to the technology that may be suitable or proper in context to a particular community, area, region or country. Appropriate technology has three dimensions that is economic, environmental and social and in a way contributes to the economic, social and environmental objectives of development having developmental goals, the resource endowment and the condition of application.

Understanding Transfer of Technology

In the above background of technology or for that matter appropriate technology, how the Transfer of Technology should be viewed. Transfer in common paralance means change of place or physical moment or mechanical transport. Then does 'TOT' mean merely taking a new idea, produce or product from extension or research institute to farmers field? Even if we take TOT literally, no idea or product can be moved just by lifting it physically. Mere physically shift will not ensure its use. People are not passive recipient to apply a technology just because it was brought to them or told to them. Thus transfer of technology cannot be led to acceptance. People have their attitude, opinions, values and established ways of doing things. New technology has to complete and work within the context and environment of people. Extent of acceptance is influenced by variety of personal, as well as social, environmental and resource related factors. Acceptance demand new learning and adjustment. Thus here lies the role of Television channel that through constant repetition of the ideas/technology through various form over a period of time as frequently as possible to widely dispersed heterogeneous audience and gradually developing or steering their mind in favour of the technology through highlighting the merits of the new ones over the conventional one which is not ordinary possible for grass-roots extension functionaries to deal with vast majority on interpersonal mode.

Required Intervention for A Greater Cause

Thus it is in this context and background, a dedicated channel exclusively meant for agriculture and allied sector on 24*7 (24 hrs. in a day and all seven days in week) mode which is practically 8 hrs. fresh programme to be repeated thrice (3 times) a day for the benefit of potential farmers viewers of the country in general and our state Odisha in particular, in regional language of the state, in our case it is 'ODIYA' language for our state. Such type of agriculture based TV will thus contribute immensely for the growth and development of agricultural sector both across the country as well as the state.

Contents to be covered by the dedicated channel

The new channel will cover all aspects of agricultural development in the state. like:

A. Agricultural Developments programme based on

1. Crop production
2. Crop improvement

3. Crop protection
4. Soil science and soil health
5. Agricultural engineering and mechanized farming
6. Food preservation, protection and processing
7. Agricultural Business management and marketing
8. Day to day meteorological information and tracking cyclonic movements

B. Horticultural Development Covering

1. Plantation crops like: Growing of orange, apple pomegranate, pineapple, citrus, papaya, banana, custard apple, grapes etc., their harvesting, grading preservation, B. packaging and export.
2. Floriculture: Growing of various flowers, plucking, preservation, packaging, extraction of essence etc. and exports.
3. Vegetables: Modern methods of growing of vegetables, harvesting, preservation, packaging, marketing, food preservation and exports process.
4. Gardening: Modern methods of designing, planning and maintenance of community as well as household gardens, kitchen gardens and creating a pollution free living atmosphere.

C. Animal Sciences

1. Animal husbandry,
2. Animal nutrition, health and sanitation,
3. Animal rearing, production and processing of healthy animal products,
4. Healthy rearing of goats, sheep, bullocks, cows, rabbits, poultry and other edible birds,
5. Food processing, marketing and export.

D. Fisheries Sciences

1. Modern scientific methods of fish farming,
2. Rearing, catching, packaging, preservation and export of fish products,
3. Fish food, health and sanitation of fish culture,
4. Fish food preparation, preservation, processing, packaging and exports.

E. Home Science/ Community Science

This segment/slot of the programme will show the very imparting of knowledge and training to the house wives on the methods of healthy food preparation, preservation, processing, packaging and if possible exports and business managements for self sufficiency. Imparting scientific knowledge on the preparation of jam, jelly, marmalades, squash, sauce, pickles, papads, badies etc. will help the unemployed house wives to be self-sufficient in their earnings through the self-help groups formed in village levels.

F. Other areas such as Forestry, Soil Conservation, Agricultural Product Market Management can also be included in the list.

Consequent Benefits

1. All these subjects will be covered/telecast in state language through a TV channel help farmers improve their knowledge in different sections of agricultural production.
2. The TV channel may be adopted by the state government or by private sectors under the control of the government of Odisha.
3. The corresponding specialist/agricultural experts working in Agricultural Universities and State Departments will information/expertise for the purpose. assist in providing specialized scientific
4. It will be a source of encouragement for the farmers to produce more food in scientific ways and enhance their economic standard as well as economic status of the state and lead towards self-sufficiency in food production.
5. It will also provide self-employment for unemployed youth of the rural areas as well as engagements for labour class.
6. The experts in various branches of agricultural science will get opportunity to understand the problems of farmers and solve them in modern scientific manner.
7. The farmers will gain day to day information on the scientific agricultural practices, marketing, source of material ete. through wide scale advertisements.
8. The Government will reach the lowest level of the rural citizens to highlight the developmental activities relating to agricultures through TV programs.

9. The farmers will be enlightened with the benefits of land utilization, water resources, crop insurances, consolidation of holdings, provisions on subsidies to farmers.

Forms/Types of Telecast Will Be

- Quiz competition with different institutions imparting knowledge/skills or conducting training on agriculture and allied sector.
- Panel Discussion on emerging issues directly or indirectly linked with agriculture. Live telecasting of success stories of different localities of different agroclimatic zones.
- Chitchat with eminent personality from field of agriculture or from farming community.
- Highlighting grass-roots innovations.
- Folk dance, Folk Art, Folk Songs based on agricultural theme.
- Question answer session relating to any topic from agriculture and allied sector on seasonal basis.
- Invited guest /resource person / master trainer/ agro entrepreneur from among farming community for sharing experience with regard to their operational constraints, marketing constraints etc.
- Opinion/Observation/Remarks of student friends from different agricultural institution including their vision, mission and strategies for future growth and development in agriculture and allied sector.
- Highlights the problems of different stake holders linked with agricultural production and productivity.
- Feedback of input agencies, financial institutions, and administrators, top level extension managers, scientists, grass roots level extension functionaries, NGOS, Elite mass speaking about the efficacy of various programmes directed towards the growth and development of agriculture and allied sectors as implemented by both central government as well as state government for the purpose.

However following parameters are to be considered before commissioning the project.

1. Technical Feasibility: All ABC relating to the development of infrastructures, and the help of Camera Person, Audio Specialist, Sound Engineers, Electrical Engineers, Software Engineer/IT Specialists,

Designer of Graphic Aids, Animation Specialist, Air-Conditioning Engineers, will be involved to judge the technicality of the TV channel for its effective and efficient functioning, in addition to script writer, programme developer and programme executive to ratify the software as well as hardware dimension of the technology focusing upon validating the contents of the programme a panel of subject experts before the said programme is telecasted.

2. Financial Sustainability: It will be basically depending upon the amount to be charged for the commercial advertisement made by seed company, fertilizer company, pesticides company, farm machinery industries and banking sector in addition to insurance company in the dedicated channel which may take time to gain momentum. A bench mark study in this regard has to be done before going ahead.
3. Social Acceptability: The very social structure, social system, social institution and social organization will be given due importance while the programme will be developed so that it will be well appreciated by the society in general and agrarian society in particular.
4. Cultural Compatibility: Looking into the food habit, religious bent of mind, taboos, mores, folkways, tradition, customs, programme will be drafted to fit well with the situation. More particularly the language matters a lot in this programme which will by odiya only for Odisha state.
5. Legal Tenability: Following all the protocol, administrative approval from the government and other concerned authorities will be obtained to make it legally more tenable.

Conclusion

Though the proposal of establishing a dedicated channel appear to be bit difficult but definitely not impossible to ground it or commission it in real sense. Though a strong will power with a positive attitude on the part of an individual or a group steering this venture backed by political will power through government support we can comfortably transfer the mental acceptance of the idea to its physical action of all stakeholders which need the very cooperation and support from within at initial stage before soliciting the same from outside to endorse the statement "Charity Begins at Home". Thereafter like other channel, it will be available in the YouTube which will be accessible to the farmers friend their cell phone/ smart phone which they always keep with them inside their pocket as a gesture of their activeness in social media. The take up point in this respect is some institution/individual/group or media person should take initiative to bell the cat, then things will follow in sequence.

Acknowledgement

The authors through this write up express their gratitude as well as indebtness to the important personality Dr. Parsuram Nayak, thệ Former Principal Scientists, Plant Pathology, National Rice Research Institute ICAR, Cuttack who mooted the idea of dedicated TV channel in odiya exclusively for Agriculture and Allied Sectors and Dr. V.V. Sadamate, Former Advisor to Planning Commission, Govt. of India who encouraged, appreciate, and gave some valuable suggestions while responding to this issue in the capacity of Chairman of the technical session “Capacity Building and Transfer of Technology” during National e-conference on Future Strategic Technologies Innovations and Entrepreneurship Development on Agriculture and Allied Sector for Food and Livelihood secretary organized by S’O’A (Deemed to be University) from 14”-16” September, 2020.

18

Assessment of Factors Affecting Adoption of Potato Production Technology in Odisha

I.P. Pattnaik, S. Jena and A.P. Kanungo

Introduction

Potato (*Solanum tuberosum* L) is one of the major starchy tuber vegetable crops of the world. It is a unique crop which can supplement the food needs of the country in a substantial manner. Today, potato is the need of life or poor man's food in many countries including India. Potato production in India has made tremendous progress only with the development of new technologies in different fields of agriculture. It has been observed that the farmers are not adopting recommended practices and consequently not realizing the potential yield of potato. To understand the problems of farmers in the adoption of complete recommended farm practices, researchers as well as extension workers should come out with some concrete findings and suggestions, which can solve these problems. It has been seen that a large number of techniques and practices do not reach the farmer's fields and those carried to the farmers get considerably distorted or often adopted partially with the result that farmers do not get the potential yield. The problem of technological change is not as simple as it is commonly understood. Therefore, it is very essential that the farmers must be aware about recommended farm practices for increasing the potato production. This study aims to find out socio-economic profile, knowledge and adoption level, factors affecting potato production technologies of farmers and constraints perceived by them in production.

Materials and Methods

The study was conducted in Nimapada and Pipli block of Puri District, Odisha in the year 2018-19. Both Purposive and Random sampling procedures were followed for selection of district, block, gram panchayat, village and respondent. The total sample size of the study was 100. The response was obtained from each individual respondent through a pre-tested structured interview schedule. The data collected was tabulated and analyzed with the help of suitable statistical tools and techniques such as frequency percentage, mean, rank, gap analysis, co-variance, correlation coefficient, multiple regression, test of significance, factor analysis etc.

Results and Discussion

It is revealed from Table 1 that majority (65%) of the respondents were in the category of middle age group that is 36-55 years and 72 per cent had higher education level. Majority (52%) of them were small farmers and maximum (84%) farmers had medium farm power possession. Majority (55%) farmers had medium level of annual income while 83 per cent of the respondents had medium level of scientific aspiration. 75 per cent farmers had medium level of membership in different social organization. The farm broadcast through radio programme was the major source of farm information which ranked 1st with highest mean score of 3.26 followed by farm telecast and newspaper. The major findings of the study were in line with the findings of Kumar A (2007), Khalil *et al* (2013) and Ahmad *et al* (2014)

Table 1: Socio economic profile of potato cultivators

Categories	Frequency	Percentage	Mean Score	Standard Deviation
Age				
Young age	7	7.0	2.210	0.556
Middle age	65	65.00		
Old age	28	28.00		
Education				
Low	5	5.0	3.390	1.163
Medium	23	23.00		
High	72	72.00		
Land Holding				
Marginal	37	37.0	1.78	0.746
Small	52	52.0		
Medium	7	7.0		
Large	4	4.0		
Annual Income				
Low	39	39.00	1.690	0.662
Medium	55	55.00		
High	06	6.0		
Farm power possession				
Low	09	9.0	4.16	1.204
Medium	84	84.00		
High	07	7.0		
Scientific Aspiration				

Categories	Frequency	Percentage	Mean Score	Standard Deviation
Low	08	8.0	15.58	2.425
Medium	83	83.00		
High	09	9.0		
Membership in Social Organisation				
Low	19	19.00	10.10	2.410
Medium	75	75.00		
High	06	6.0		
Sources of Farm Information	Mean Score	Rank Order		
Newspaper	3.21	III		
Farm broadcast	3.26	I		
Farm telecast	3.24	II		
Personal contact	3.19	IV		
Meetings/Training	1.85	V		

Table 2 revealed that age (25.152%), membership in social organization (23.861%), media exposure (13.335%) and scientific aspiration (15.567%) of the respondents had higher consistency than other socio economic variables of the respondents. At the same time education, land holding size, farm power possession and annual income of the respondents had a greater variability. From the findings it can be concluded that age, membership in social organization, media exposure and scientific aspiration of the respondents were the important socio economic variables which may be taken into consideration for gain in knowledge and promoting adoption of technology. The major findings of the study were in line with the findings of Kumar A (2007) and Khalil *et al* (2013).

Table 2: Covariance analysis of socio-economic variables

Sl. No.	Variables	Mean	S.D.	S.E.	CV%
1	Age	2.210	0.556	0.056	25.152
2	Education	3.390	1.163	0.116	34.292
3	Land holding size	1.780	0.746	0.075	41.935
4	Average annual Income	1.690	0.662	0.066	39.173
5	Farm power possession	4.160	1.204	0.120	28.935
6	Scientific Aspiration	15.580	2.425	0.243	15.567
7	Membership of Social Organization	10.100	2.410	0.241	23.861
8	Media exposure	30.890	4.119	0.412	13.335

A perusal of table 3 depicted that the respondent farmers had sound knowledge in intercultural operation with highest mean score of 2.96, followed by plant protection measures (2.94), variety (2.92) and soil & land preparation (2.88) whereas they had satisfactory knowledge on nutrient management (2.67). But they had somewhat poor knowledge on planting (2.32) and harvesting and post harvest technology (2.23) of potato production. With regard to knowledge gap of respondents on potato production, it was revealed that harvesting and post harvest management practices ranked first. Planting, nutrient management, soil and land preparation, variety and plant protection measures ranked second, third, fourth and fifth respectively in order of importance. The lowest knowledge gap (1.33%) was observed in case of intercultural operation. It can be concluded that higher the knowledge level (mean score), lower is the gap in knowledge of the respondents. The major findings of the study were in line with the findings of Khalil *et al* (2013) and Ahmad *et al* (2014).

Table 3: Knowledge level of farmers on major areas of potato production technologies

(N=100)

Knowledge level	**Fully known (3)**		**Partially known (2)**		**Not known (1)**		**Mean score**	**Rank**	**Gap%**
	F	**%**	**f**	**%**	**F**	**%**			
Soil and land preparation	**88**	**88.0**	**12**	**12.0**	**0**	**0**	**2.88**	**IV**	**4.00**
Variety	92	92.0	8	8.0	0	0	2.92	III	2.67
Planting	32	32.0	68	68.0	0	0	2.32	VI	22.67
Intercultural operations	96	96.0	4	4.0	0	0	2.96	I	1.33
Nutrient management	67	67.0	33	33.0	0	0	2.67	V	11.00
Plant protection measures	94	94.0	6	6.0	0	0	2.94	II	2.00
Harvesting and post harvest	23	23.0	77	77.0	0	0	2.23	VII	25.67

It is revealed from table 4 that the respondent farmers had an adoption gap of 30.33 percent in case of use of proper potato varieties which ranked 1st, followed by a gap of 29.67 percent (2nd), 25.67 percent (3rd), 25.33 percent (4th), 6.67 percent (5th), 4.00 percent (6th) and 2.67 percent (7th) in the broad areas of planting, harvesting and post harvest, nutrient management, intercultural operations, soil and land preparation and plant protection measures. The higher production gap in some areas of adoption might be due to the fact that lack of awareness, lack of exposure to latest technologies, less extension contact and inadequate training . The factor analysis of adoption level of respondents revealed that two major areas of potato production technologies such as soil and land preparation and selection of variety exerted considerable impact upon the total adoption level of the potato growers. This was because of their long experience and refinement in years of potato production as well as knowledge on improved and mechanised farming. The major findings of the study were in line with the findings of Kumar A (2007) and Ahmad *et al* (2014).

Table 4: Adoption level of farmers on major areas of potato production technologies

(N=100)

Sl. No.	Major areas of adoption	Fully Adopted (3)		Partially Adopted (2)		Not Adopted (1)		Mean Score	Gap%	Rank
		F	%	f	%	F	%			
1	Soil and land preparation	88	88.0	12	12.0	0	0	2.88	4.00	VI
2	Variety	9	9.0	91	91.0	0	0	2.09	30.33	I
3	Planting	11	11.0	89	89.0	0	0	2.11	29.67	II
4	Intercultural operations	80	80.0	20	20.0	0	0	2.80	6.67	V
5	Nutrient management	24	24.0	76	76.0	0	0	2.24	25.33	IV
6	Plant protection measures	92	92.0	8	8.0	0	0	2.92	2.67	VII
7	Harvesting and post harvest	23	23.0	77	77.0	0	0	2.23	25.67	III

It is revealed from table 5 that among the 08 variables, age and scientific aspiration have recorded significant regression impact upon knowledge of the respondents. The R^2 value being 0.313 it is to conclude that the conglomeration of causal factors together has attributed to 31.3 percent, indicating considerable influence in increasing knowledge level of the respondents. The findings are collaborated with the findings of Kumar A (2007) and Khalil *et al* (2013) It is observed From table 6 that a good number of socio-economic variables such as: media exposure, land holding size, farm power possession, annual income and scientific aspiration had very significant impact upon the adoption level of the respondents regarding various production related technologies. The major findings were in line with the findings of Kumar A (2007), Khalil *et al* (2013) and Ahmad *et al* (2014).

Table 5: Multiple regression analysis of socio economic variables on knowledge level of respondents

Coefficients[a]						
Model		Unstandardized coefficients		Standardized coefficients	't'	Sig.
		B	S.E.	Beta		
1	(Constant)	14.264	1.476		9.667	0.000
2	Age	0.942	0.265	0.426	3.558	0.001**
3	Education	0.112	0.133	0.106	0.842	0.402
4	Land holding size	0.023	0.195	0.014	0.119	0.905
5	Avg. Annual income	0.010	0.238	0.005	0.041	0.968
6	Farm power	0.108	0.161	0.106	0.674	0.502
7	Scientific aspiration	0.196	0.117	0.387	1.675	0.098**

Coefficients[a]						
Model		**Unstandardized coefficients**		**Standardized coefficients**	**'t'**	**Sig.**
		B	**S.E.**	**Beta**		
8	Membership of social organization	-0.070	0.072	-0.138	-0.976	0.332
9	Media exposure	-0.061	0.052	-0.205	-1.170	0.245

Model summary				
Model	R	R Square	Adjusted R Square	Std. Error of the Estimate
1	0.559a	0.313	0.209	1.09235

a. Predictors: (Constant), Age, Education, Land holding, Annual income, Farm power, Scientific aspiration, Membership of social organization, Media exposure

Table 6: Correlation study of socio-economic variables with adoption level (N=100)

Sl.No.	Variables	r-value	t-calculated	t-tabulated (df 98)
1	Age	0.091	0.905	2.627 At 0.01 level of significance 1.984 At 0.05 level of significance
2	Education	0.180	1.812	
3	Land Holding size	0.211*	2.137*	
4	Avg. annual Income	0.409**	4.437**	
5	Farm power	0.398**	4.295**	
6	Scientific Aspiration	0.545**	6.435**	
7	Membership of Social organization	0.075	0.745	
8	Media exposure	0.437**	4.809**	

**Correlation is significant at the 0.01 level (2-tailed).

*Correlation is significant at the 0.05 level (2-tailed).

Table 7 depicted that out of 8 variables such as- age, annual income and scientific aspiration had recorded significant regressional impact upon adoption level of the respondents. The R^2 value being 0.385, it is to conclude that the conglomeration of causal factors together has attributed to 38.5 percent, so it can be concluded that the socioeconomic variables had considerable influence in accelerating adoption level of the respondents. The major findings were in line with the findings of Kumar A (2007) and Khalil *et al* (2013)

Table 7: Multiple regression analysis of socio economic variables on adoption

Coefficients[a]						
Model		**Unstandardized coefficients**		**Standardized coefficients**	**t**	**Sig.**
		B	**S.E.**	**Beta**		
1	(Constant)	9.521	1.669		5.704	0.000
	Age	0.787	0.300	0.298	2.626	0.010**
	Education	0.154	0.151	0.122	1.022	0.310
	Land holding size	-0.189	0.221	-0.096	-0.855	0.395
	Avg. Annual income	0.354	0.269	0.159	1.316	0.192*
	Farm power	-0.215	0.182	-0.176	-1.181	0.241
	Scientific aspiration	0.385	0.132	0.636	2.908	0.005**
	Membership of social organization	-0.038	0.082	-0.063	-0.467	0.641
	Media exposure	0.017	0.059	0.048	0.290	0.773

Model summary				
Model	R	R Square	Adjusted R Square	Std. Error of the Estimate
1	0.621[a]	0.385	0.293	1.23582

a. Predictors: (Constant,), Age, Education, Land holding, Annual income, Farm power, Scientific aspiration, Membership of social organization, Media exposure

A perusal of table 8 depicted that out of five major broad areas of constraints, the marketing constraint was the top most constraint perceived by the farmers and ranked 1st with mean score of 14.46, followed by technical constraints at 2nd rank, management related constraints at 3rd rank, economic constraints at 4th and production related constraints at 5th rank with mean score of 10.60, 10.29, 9.38 and 9.15 respectively. Therefore, proper attention should be given to eradicate the major constraints of the potato growers of Odisha giving emphasis on the marketing, production technology and management related constraints to boost potato production and fetch remunerative price for the farmers produce. Majority of the findings were in line with the findings of Kumar A (2007) and Ahmad *et al* (2014).

Table 8: Comparative analysis of different constraints experienced by the respondents

Sl. No.	Constraints	Mean score	Rank
1	Production related constraints	9.15	V
2	Management related constraints	10.29	III
3	Technical constraints	10.60	II
4	Economic constraints	9.38	IV
5	Marketing constraints	14.46	I

Conclusion

Potato has a great demand irrespective of the rich and poor. There is a wide gap between present production and consumption levels of potato in our state Odisha. In order to fulfill the requirement of the consumers of the state and to fetch potato producers a remunerative price, timely supply of quality seed materials and production inputs, construction of more number cold storages, godowns, strengthening producer-market linkage, need based trainings, effective monitoring and supervision and reducing the communication gap between extension functionaries and farmers is of paramount importance. Therefore the policy makers, planners, scientists may take necessary steps to train, motivate and enhance the knowledge and adoption levels of the farmers regarding modern potato production technologies in order to bring desirable change in the potato production scenario of Odisha state.

References

Ahmad Peer QJ, Dar MA, Malik HA and Kaur J. 2014. Multiple regression analysis for adoption studies of potato growers in Jammu division, Journal of Applied and Natural Science, 6(2): 664 – 671.

Benal D, Patel M, Jain MP and Singh VB. 2010. Adoption of Dryland Technology, Indian Journal of Dryland Agriculture Research& Development, 25(1): 111-116.

Khalil MI, Haque ME and Haque MZ. 2013. Adoption of BARI Recommended Potato (Solanum tuberosum) Varieties by the Potato Farmers of Bangladesh, The Agriculturists-A Scientific Journal of Krishi Foundation, 11(2): 79-86.

Kumar A. 2007. Farmers' perception of the attributes of selected potato technologies, Potato Journal, Central Potato Research Institute, Simla, 34(1-2): 141-142.

Sharif Uddin ABM, Rahman MM, Alam MB and Kamaly MHK.2014. Factor analysis towards perception of potato production technologies by potato growers in Bangladesh, Journal of Global Communication, 7(2): 97-104.

19

Increasing Income of Farmers and Improving the Soil Fertility Level by Way of Soil Health Card

A. Tiwari

Introduction

Soil is one of the elements required for farming and it provides nutrients to the plant. Healthy soil containing all the elements for growth and development of crop as well as heart of soil health is the integration of the soil physical, chemical, and biological process and its functions. But inorganic farming adopted over the recent years to maximize crop production has become Hazardous because in the long run deteriorating the soil health and the quality of the agricultural produces besides polluting the environment and natural resources of soil and water dangerously. Soil testing helps to recommend chemicals fertilizers more judiciously in combination with organic manures and bio fertilizers and hence balanced nutrition to crop. Soil health and fertility is the basis for sustainable profitability of the farmers. Soil testing is a science based and time-tested tool for assessment of soil fertility status and soil ailments and for nutrient amendment recommendations. The current consumption of NPK ratio is 6.7:2.4:1 which is highly skewed towards nitrogen as against ideal ratio of 4:2:1 in India. Hence, government of India introduced Soil Health Card Scheme on 19 February, 2015 across India after assessing urgent need for balanced use of fertilizers. The soil health card is complete evaluation of the quality of soil properties. It contains corrective measures that a farmer should adopt to obtain a better yield, and increase your incomes with increasing productivity. Keeping above problematic condition, the present study conducted to investigate farmer's income and level of soil fertility level.

Material and Methods

The study was based on primary data as well as secondary data and multi-stage random sampling method was used to select farmers. Two blocks were selected randomly from 16 blocks of Deoria district and thereafter one village was selected randomly from each block. A list of farmers from each village were prepared and categorized into soil card holders and non-soil card holders. The 15 farmers were randomly selected from each category and each village and made the sample of 60 farmers. The descriptive analysis (percentage) and CACP (Commission for Agricultural Costs and Prices) cost concepts were used to estimate the impact of soil health card on farmers' income.

CACP cost concepts

Cost A1= wages of hired human labour (in Rs.) + value of bullock labour (in Rs.) + value of tractor labour (in Rs.) + value of seeds (in Rs.) + value of manures and fertilizers (in Rs.) + value of plant protection chemicals (in Rs.) +value of irrigation charges (in Rs.) + value of land revenue (in Rs.) + value of depreciation on fixed capital assets (in Rs.) + value of interest on working capital (in Rs.)

Cost A2= Cost A1 + rent paid for leased-in land

Cost B1= Cost A1 + interest on fixed capital (excluding land)

Cost B2= Cost B1 + Rental value of owned land + Rent for leased-in land Cost C1= Cost B1 + Imputed value of family labour

Cost C2= Cost B2 + imputed value of family labour

Cost C3= Cost C2 + 10 per cent of cost C2 as management cost

Result and Discussion

Table 1 shows the yield of rice, wheat, barley, mustard, potato and sugarcane were increased with 33.31 per cent, 26.77 per cent, 6.88 per cent, 33.40 per cent, 49.29 per cent and 16.29 per cent respectively at the farm of soil health card holder over the non-soil health card holders.

Table 2 shows the cost of cultivation cost per hectare of Rice, Barley, Sugarcane, Potato, Wheat and Mustard for soil health card holder is Rs 22589, Rs 19858, Rs 53589, Rs 44987, Rs 27898 and Rs 22158 respectively and for non-soil health card holder is Rs 24659, Rs 21173, Rs 55651, Rs 46931, Rs 29813 and Rs 23147 respectively. The cost of cultivation recorded highest decreased in rice (-8.39 per cent) and lowest in sugarcane (-3.70 per cent) for soil health card holder as compare to non-soil health cardholder.

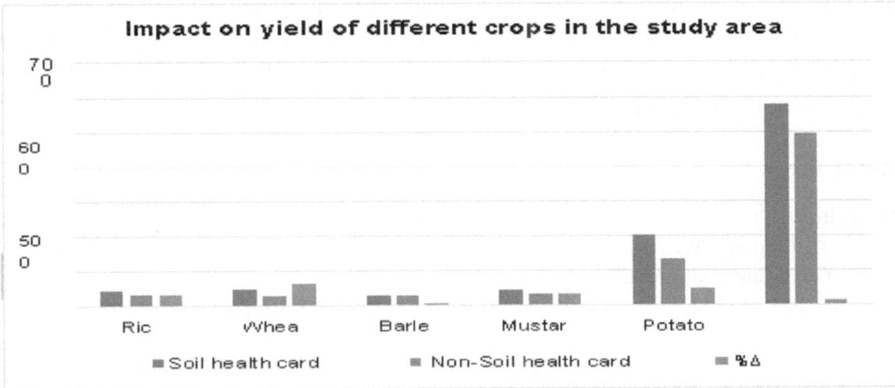

Table 1: Impact on yield of different crops in the study area (in Q/ha)

Crop	Soil health card holder	Non-Soil health card holder	%Δ
Rice	46.50	34.88	33.31
Wheat	39.30	31.00	26.77
Barley	31.00	29.06	6.88
Mustard	6.51	4.88	33.40
Potato	201.55	135.00	49.29
Sugarcane	575.66	495	16.29

Table 2: Cost of different crops by SHCH* Farmer and Non-SHCH** farmer (in Rs. /ha)

Crops	Cost A			Cost B			Cost C		
	SHCH	NSHCH	%Δ	SHCH	NSHCH	%Δ	SHCH		%Δ
Rice	17790	18055	-1.46	20585	21651	-4.92	22589	24659	-8.39
Barley	13932	14525	-4.08	16584	17965	-7.68	19858	21173	-6.21
Sugarcane	34525	36671	-5.85	43589	45365	-3.91	53589	55651	-3.70
Potato	31258	33176	-5.78	36589	38958	-6.08	44987	46931	-4.14
Wheat	20582	21354	-3.61	24595	26197	-6.11	27898	29813	-6.42
Mustard	17528	18417	-4.82	20890	21897	-4.59	22158	23147	-4.27

*SHCH- Soil Health Card Holder and **NSHCH-Non-Soil Health Card Holder

Conclusion

The study has represented that adoption of soil health card leads to increased yield of different crops due to using of recommenced doses of fertilizers because most of the farmers were used lower quantity of fertilizers. The cost of cultivation of different crops decreased and as well as yield increased in case of soil health card holder as compare to non-soil health card holder. The advantages of adoption of soil health card are disseminated among the farmers so that farmers were benefited from this scheme.

References

Agriculture at a Glance (2018). Directorate of Economics and Statistics, Ministry of Agriculture & Farmers Welfare, Department of Agriculture, Cooperation &Farmers Welfare, Government of India.

R. S Chauhan, H.O Sharma, D. Rathi and H.K. Niranjan (2017), Impact of Soil Health Card on Farmer's Income. Agricultural Economics Research Review, 30.139-141.

Susheela Meena, I.P. Singh and Ramji Lal Meena (2016), Cost of cultivation and returns on different cost concept basis of onion in Rajasthan. Economics Affairs, 61(1). 11-16.

20

A Study on Training Need Assessment of Coconut Growers

M. Sahoo, I. Pattanaik and S.K. Rout

Introduction

For capacity building of farming community, training plays a vital role. It helps in enriching the knowledge along with enhancement of essential skill required in performing certain practices in a specific crop cultivation. The skills required by a coconut cultivator need to be supplemented with additional skills in order to increase the yield of nuts. Government provide better support to the coconut growing farmers to improve their income, to reduce the risk involved in cultivation, implementation of quality agricultural practices and finally enhancing the income of coconut grower in business mode. Training is a source of information, advice and influence in the decision making process. Training helps in enhancement of knowledge, improvement of skill, interaction with experts and to build confidence. Keeping this in view, the present study was conducted to know the training need areas of coconut growers in Gop block of Puri district of the state of Odisha.

Materials and Methods

The study was conducted during the year 2015-16 to assess the training need of coconut growers and to study the personal and socio- economic profile, irrigation needs, opinion and suggestion of coconut growers with respect to place, time, duration and method of training. Three villages from Gop block of Puri districts, namely Siripur, Anasara and Junei were selected purposively where the DBT- RBC project was operated by OUAT. The respondents were selected on random sampling technique of coconut growers were selected for the study. The total sampling was 180 from the three villages (60 from each village).

The areas of training need that were consider of the study: land preparation, seed rate, selection of quality seeds, nursery management, age of the seedling, spacing, manures and fertilizer of the management, intercropping, weed management, pest management, disease management that includes preventive, cultural, mechanical, biological and indigenous harvesting and marketing. In addition to that the suggestion and opinion was collected with regards to place of training(village level, panchayat level, district level, KVKs), time of training (April-May, June-July, September-October, December-January), duration of training (one day, two days, five days, one week), method of training (lecture, lecture +discussion, lecture + discussion + method demonstration, lecture + discussion + method demonstration + field visit).

The personal and socio economic characteristic considered were age (young, middle, old) , educational qualification (illiterate, middle english, high school, graduate), family size (nuclear, semi- nuclear, joint), land holding (small, large), farming operation (little, medium, high), extent of exposure (localite, semi-cosmopolite, cosmopolite) With respect to the area of training the respondents were collect in an interview schedule on three point scale like very important, important and not important. The data was analyse by assigning three, two and one score respectively. Mean score of each training area were calculated by using the formula:

$$\text{Mean Score} = \frac{\text{Very important* 3 + important *2 + not important *1}}{\text{Total number of very important + important + not important}}$$

Results and Discussions

The data presented in table 1, shows that majority (46.7%) of coconut growers belongs to middle aged category followed by (28.9%) and (24.4%) old aged category and young aged category respectively. Usually the middle aged coconut growers are highly enthusiastic, having more responsibility and have more work efficiency than the younger and older one.

Table 1: Personal and socio-economic profile

S. No.	Particurals	Characterstics	No. of respondents	Percentage (%)
1	Age	Young	44	24.4
		Middle	84	46.7
		Old	52	28.9
2	Educational qualification	Illiterate	16	8.9
		Middle english	34	18.9
		High school	86	47.8
		Graduation	26	14.4
		Above	18	10
3	Family size	Nuclear	106	58.9
		Semi- nuclear	50	27.8
		Joint	24	13.3

S. No.	Particurals	Characterstics	No. of respondents	Percentage (%)
4	Land holding	Small	104	57.8
		Large	76	42.2
5	Farming operation	Little	54	30
		Medium	76	42.2
		High	50	27.8
6	Extent of exposure	Localite	122	67.8
		Semi- cosmopolite	44	24.4
		Cosmopolite	14	7.8

With respect to education level of coconut growers, majority (47.8%) studied up to high school followed by middle english, graduation, above graduation level and illiterate. Interestingly, it was observed that very negligible per cent of people were illiterate. Majority (58.9%) of coconut growers belonged to nuclear family type followed by (27.8%) and (13.3%) belongs to semi nuclear and joint family respectively. This may be due to awareness of family planning, to provide better education to their children and to have a better standard of living. As regards to the land holding it is cleared from the table-1 that (57.8%) belonged to small farmers and (42.2%) belongs to large farmers.

With respect to the farming experience or operation farmers are almost equal experienced in different category and they are having evenly distributed. This situation may be because of the exposure and their involvement in coconut cultivation. Further, (67.8%) of the farmers have localite exposure followed by (24.4%) cosmopolite exposure and (7.8%) semi-cosmopolite exposure. This could be due to majority of the respondent possessing their exposure at their locality level due to coconut cultivation is very popular in those areas. With regards to the training needs of coconut growers (table 2) among several needs, majority of coconut growers expressed their needs with respect to disease management and pest management. Those were the major training areas with high mean score of (173.53) followed by (95.07) and (68.30).

Table 2: Areas of irrigation needs.

Sl.No	Areas	Very important	Important	Not important
1	Land preparation	18	48	118
2	Seed management	134	34	12
3	Nursery management	116	72	22
4	Age of plant	22	70	88
5	Spacing	24	62	94
6	Manure and fertilizer	132	28	20
7	Intercropping	90	58	32

Sl.No	Areas	Very important	Important	Not important
8	Water management	158	14	8
9	Pest management	180	0	0
10	Disease management	180	0	0
11	Harvesting time	0	0	90
12	Marketing	74	58	48
	Mean	173.53	68.30	95.07

Disease and pest management are the major area where the training are very important as expressed by majority of the respondent i.e: cent %. followed by (87.78%),(74.44%),(73.33%),(64.44%),(50%),(41.11%),(13.33%),(12.22%) and (10%). The results presented in table 3 shows that the opinion of coconut growers towards place, time, duration and method of training with respect to place of training KVK/RRS was the place opted by majority(85.56%) of the respondents followed by panchayat and village (7.78%), (5.56%) respectively. The probable reason for selection of KVK or RRK case as a place for training might be due to KVK equipped with full training facility technically sound and well qualified faculty and comfortable accommodation facility for staying for the participants in the training period. April to June was suitable training time as opted by cent % of the coconut growers.

Table 3: Opinion on irrigation management

Sl. No	Particulars	Characteristic	No. of Respondents	Percentage (%)
1	Place of training	Village	12	6.7
		Panchayat	14	7.8
		Districts	0	0
		KVK	154	85.5
2	Time of training	April,may	92	51.1
		June, july	50	27.8
		September, October	18	10
		December, january	20	11.1
3	Duration of training	1 day	46	25.6
		2 days	134	74.4
		5 days	0	0
		One week	0	0

Sl. No	Particulars	Characteristic	No. of Respondents	Percentage (%)
4	Method of training	Lecture	0	0
		Lecture + Discussion	77	42.8
		Lecture+ Discussion +Method demonstration	63	35
		Lecture + discussion+ method demostration+ field visit	40	22.2

With regards to duration of training 2 days are the ideal duration as opted by majority of the respondents (74.44%) followed by one day (25.56%). This could be due to non availability of farm labourers as they get engaged in one or other farming activities. Agricultural practices should be taken up at right time in order to avoid future consequences otherwise this may leads to so many problem. This is the reason for which they have opted for short duration training programme rather than long duration.

Conclusion

It is concluded from the study that majority of the coconut growers expressed that training needs regarding disease management and pest management followed by water management, selection of quality planting material, manures and fertilizer management and for duration and month place of training and 2 days duration training programme in the month of April to July. The venue have been opted as KVK/RRTTS and a combination of training programme has been chosen by the majority of the farmers because they develop confidence through the combination of training along with visit and exposures. Hence, the policy makers and administrators have to make a suitable provision for intensive training programme for coconut growers as per the training needs of the coconut growers. Therefore the developmental departments and extension organizations involve in the rural development and extension activities should plan for suitable training programme to have effectiveness of the training and its use by the coconut growers with ultimate objective of their enhancement of income.

References

Borich, G.D., 1980. A needs assessment model for conducting follow-up studies. J. Teach. Educ. 31: 39-40.

Hashemi, S.M., Mokhtarnia, M., Erbaugh, J.M.,Asadi,A.,2008.Potential of extension Cord 2016, 32 (2) 20 and awareness of IPM. Sci. Total Environ. 407: 84- 88.

Trutmann, P., Voss, J., Fairhead, J., 1996. Local knowledge and farmer perceptions of bean diseases in the Central African Highlands. Agr. Human Values 13: 64-70.

21

Organic Pineapple Production Technology in Tripura: The Lone AEZ for Fruits in North East India

U. Das, R. K. Bhattacharyya, P. Bhattacharjee and P. Choudhury

Introduction

The hilly states in north eastern India are not only famous for their forest wealth and awesome wild assets. These states are also known for their track record in harvesting specific crops and fruits. Some of the iconic ones are like, "Assam tea" which is famous in and outside India and popularity of "Khasi mandarin" is also well known. The small landlocked state Tripura produces quality pineapples in abundant amounts, where cv. Queen is popular among the growers. Sikkim and other states of north eastern region of India is known for their Organic Farming techniques, which is mostly attributed the nature of hill farming on this region. North-East is rich in horticultural aspect viz., floriculture with a rich variety of medicinal plants grown there along with various fruits and vegetables. An analysis of NSSO survey data on cultivation practices shows that the North-East hill states have retained traditional practices and show an inclination towards organic agriculture that is harnessed for the development of the region with ecological benefits. Use of chemical fertilizers in agriculture enhances yields but also undermines the quality of life through possible contamination of soil, water and air and even the final products that retain their residues. Such activities in hilly regions can have external effects on other regions as well.

The pineapple (*Ananas comosus* L.) is one of the leading commercial fruit crops of the tropics. Pineapple is also a commercially important fruit crop of India with around 105,000 ha area under this crop, 17.29 lakh tonnes annual production and 16.46 tonnes /ha productivity (Anonymous, 2019). Tripura stands forth position in India in terms of area (12.57 thousand ha) and fifth in production (178.45 thousand tonnes) after West Bengal, Assam, Kerala, Karnataka (NHB, 2017). Pineapple is one of the choicest fruit all over the world because of its pleasant taste and flavour. It is a good source of vitamin A and B and fairly rich in vitamin C and minerals like calcium, magnesium, potassium and iron. It is also a source of bromelin, a digestive enzyme. In addition to being eaten fresh, the fruit can also be canned and processed in to different forms.

Tripura Queen pineapple is characterized by its distinct aroma and pleasant organoleptic qualities with a comparatively lesser fiber content. Apart from this it is almost free from any chemical residue because of the attempt of organic cultivation practices. Tripura queen pineapple fruits are spiny, golden yellow in colour and release a pleasant aroma and flavour at the ripening stage. The fruits are harvested when eyes turns yellow during mid- May to mid-July when the fruit is available. Average weight of fruit varies from 1.0 to 1.5 Kg's. It possesses all the good organoleptic qualities. Its juice is of bright yellow colour. A total soluble solid (TSS) varies from 1·0 to 14° Brix depending upon the stage of maturity and season and the pH value ranges from 4 to 4.5. The water content is from 80% to 90%. The 'sweetness" and the "unique aroma' differentiates it with the pineapples of other region of North East India. Tripura Queen pineapple is not found suitable for slices because of its smaller size and deep eyes. In order to increase the shelf life and to make pineapple juice and slices available even during off-season, canning is the most popular method. The waste parts left from canning plants, including the skin, core and ends, are used to make alcohol, vinegar and food for livestock.

Nutritional Benefits

Pineapples are nutritionally packed members of the Bromeliaceae family. This delightful tropical fruit is high in the enzyme bromelain and the antioxidant vitamin C, both of which play a major role in the body's healing process. Bromelain is a natural anti-inflammatory that has many health benefits and encourages healing. Pineapple fruit is very low in Saturated Fat, Cholesterol and Sodium. Tripura Queen Pineapples contains vitamin C, which is important for the immune and digestive systems Pineapples have anti-inflammatory effects which are good for those long hard days and those heroic sporting injuries. They contain the enzyme Bromelain which is thought to help digestion of food that we eat.

Queen as the ideal cultivar for Tripura

It is the most popular and excellent cultivar of Tripura for fresh consumption. The Queen pineapple variety produced in Tripura is not just distinctly special in comparison with other varieties grown elsewhere in India but also significantly different in quality from the same variety grown in other adjacent states of North East. It is the most popular and excellent cultivar of Tripura for fresh consumption.

The plants are dwarf in stature and compact, habit of growth and bluish green foliage. The leaves are short stiff, spiny along the margins and thickly covered with a whitish bloom in both surfaces. The fruit weighs 0.8 - 1.3 kg. Peduncleis short, fruit-lets or eyes are small, prominent, deep set. When fully mature, the fruits of Tripura Queen pineapples are spiny, golden yellow in colour and emit pleasant aroma and flavour at the ripening stage. The flesh, although less juicy than Cayenne, is crisp (less fibrous), transparent with a pleasant aroma and flavour. The Total Soluble Solids (TSS) content varies from 13 to 17.2 brix and acidity varied 0.6 - 0.8 %. The slips are 1 - 4, suckers 1 - 3 and both are smaller in size than those of Cayenne.

Although these fruits are smaller than the majority of other types of pineapple, but it charms consumers with their golden yellow pulp and delicious fragrance. The juice is bright yellow in colour and possesses very pleasant organoleptic qualities. The sweetness and unique aroma of Tripura Queen pineapple differentiates it with Queen pineapple of other region of North East India. The TSS % (Total suspended solids) content of Tripura queen pineapple is as high as 18-19% whereas Total sugar % ranges from 13-14% depending upon the stage of maturity and season. One more feature that separates these pineapples from pineapples grown in other parts of the country is their lesser fiber content.

Suitable agro-climate

One of the reasons for this success is the agro climatic conditions of Tripura, which are highly, conducive for cultivation of various horticultural crops and primarily "queen', 'Kew,' varieties of pineapple. Owning to this high climatic suitability the farmers need not to use chemical inputs for cultivation and neither is there any requirement for irrigation. The North-East region produces more than 40% of the total pineapple of the country and almost 9 - 9 5% of the produce is organic. The common varieties produced from this region (North East) are qualitatively different and is said to be among the "Best in the world as they are very sweet (high TSS) with less fiber, The Tripura queen pineapple produced in Tripura is free of any chemical residue thus making the fruits of Tripura by default Organic. The pineapple fruit are highly sensitive to direct sunlight. Both bright sunshine and total shade are harmful for the plant and deteriorates quality of fruits and therefore shade management plays an important role in cultivation. Farmers at Tripura grow the Queen pineapple under the natural shaded canopy of various trees which primarily includes

Jackfruit, Litchi and other trees. It is believed that Tripura pineapple plants grown under the shade of jackfruits produces better quality of fruits.

Package of practices for Pineapple- Climate and Soil

Pineapple is natural habitat of humid tropical region. Pineapple is an evergreen perennial plat which does not have dormancy and will continue to grow as long as environment favours. Most of the Pineapples are grown in the region between 25°N and south of equator. The most northerly location for cultivation of Pineapple appears to be Assam at latitude of 30°45'N and most southern part is Port Elizabeth in South Africa at 33° 58' altitude.

Temperature range of "18°C to 32°C" is most favourable for Pineapple Cultivation. In Tripura the summers are warm to hot (19°C to 36.8°C) and wet, while the winters are cool (8°C to 27.5°C) and comparatively dry. Plant growth decreases rapidly at mean temperatures below 15° C or above 32° C. In Tripura low temperatures occur from December to January and potentially inhibit growth during winter and Queen Cultivar can initiate reproductive development below 10° C; however fruit formation is drastically affected and will eventually impact on the harvest dates. High (above 35° C) and low (below 10°C) temperatures affect development and retard growth. Pineapple plants are most productive under dry environments where low rainfall is supplemented with irrigation in well - drained soils. Pineapple performs well in relatively low water regimes; it requires as much as 5 cm of water per month from rain and irrigation, in the subtropical areas of Tripura elevation and aspects are of particular importance in deciding the site for pineapple cultivation. Therefore most pineapple plants are planted on hill sides of escape frost. Pineapple thrives well when planted on a north - easterly aspect where they receive the maximum amount of sunlight.

Soil

Pineapple plants require sandy soil and good drainage to prevent water logging therefore purposely built raised beds on slopes are utilized. Well drained loamy soil with high organic matter and a pH of 4.5 - 6.5 with low calcium content is best for pineapple cultivation. The dark brown and reddish -brown basaltic and sandy loams are considered ideal for pineapple production. The experiences of north eastern regions shows that fruit grows well in a wide range of soils, from the new alluvium of plains to old alluvium of the sub-mountainous tracts as well as the lateritic soils of the hills so long as land is not waterlogged at any time of year. Probably, no other fruit crop is possible where most of the pineapple orchard are exists in the region

Propagation

Tripura Queen pineapple is commercially propagated by suckers in Tripura. The suckers arise and grow from buds below the ground level. The slips arise from fruit stalks. They are smaller than suckers but borne more in number per plant than suckers. The crown grows on the top of the fruit. It is the vegetative growth at the top of the fruit, attached to the central core of the fruit. Fruit stalks cut into bits known as discs can also be used for propagation. Among sucker slips and crown the first two are better for early fruiting but for planting uniformity in sizes i.e. in terms of weight is a per condition for a uniform field and crop situation.

Land preparation

An area with 30-40% slope is generally selected for Tripura Queen pineapple cultivation. In the North East Region, the most common method of initial clearing of the land for plantations is by way of slash-and-burn. No formal organic compost is added to the soil for planting. Only in cases where cow dung or other compost materials are available, they may be added to the pits that have been dug for plantation. The land is generally prepared by hoeing but in areas where the slope is not too steep, the land may be levelled by ploughing to facilitate uniform distribution of water and nutrients. Alternate crisscross rows are made using a bamboo across the slope, which helps in soil and water conservation. In most hill areas of the North East Region, the fields are not ploughed but uniform rows are demarcated either across the slopes or along the slopes where the suckers are planted at uniform spacing (in the case of mono-cultivation) or random spacing if planted along with other perennial crops such as banana, papaya, etc.

Planting system

The land should be thoroughly ploughed and pulverized to a good tilt. Tripura queen pineapple is mainly planted just at the onset or the offset of monsoon, in order to avoid heavy precipitation in the pre establishment period of the plants. Though the best time for plantation in Tripura would be may to June, but availability of suckers during this period is scarce as the fruits attaining maturity at this time and the cultivators are reluctant to for de-suckering. So in general practice planting in Tripura is done during late August to early October taking the advantage of post monsoon soil moisture.

Before planting the suckers or slips should be sun cured and fry leaf scales at the base should be removed. Suckers should be planted in 10 cm deep hole but the heart of the suckers must not be buried. Planting may be done in single row or double rows following triangular and rectangular system. The following is the spacing required for different plant population per hectare.

Organic Cultivation

The region has remarkable advantages of fertile and organically rich soils, ample rainfall and water resources, river valleys, swamps and streams and great climatic diversity supporting diverse cropping possibilities. Cultivation is based on organic manure and is accompanied to a significant extent with use of improved seeds. It is interesting to note that wastes via manure and the waste generation process through animal feed and vegetable waste are used North East region is so into organic farming that one would not come across a single fertilizer shop there. Thus the pineapple coming from this region is free of any chemical residue. In organic farming, no chemical fertilizers or pesticides are used. Normally farmers use urea to increase the weight of the pineapple, which is not used in organic farming. Because the crops cannot be treated (dipping) with insecticides/ fungicides on organic pineapple plantations, the farmer is forced to pay particular attention to the quality and origin of the shoots (diseases that can be transmitted from crop to crop). This is especially the case for shoots that have been bought. In principle, it is recommended to use shoots from the plantation itself and to work very carefully.

In the majority of organic plantations, local varieties are planted together with other crops either in agro forestry or mixed crop systems. Examples exist where Pineapples are planted as a rotation-fruit with green fallow land and other crops. The farming plan will depend upon which cultivation form is adopted (agro forestry system, mixed crops as a bottom culture, crop-rotation etc.). A pineapple monoculture is not permitted in organic farming systems. To prepare the land used for pineapple production green manuring plants like e.g. *Vigna unguiculata, Crotalaria juncea or Mucuna capitata* can be sowed prior to the pineapples.

Interculture

Earthling up is an essential operation in pineapple cultivation aimed at good anchorage to plants. It involves pushing the soil into the trench from the ridge where trench planting is a common practice. As the pineapple roots are very shallow, the plants are eventually lodged especially under conditions of flat-bed planting in heavy rainfall areas. Lodging of plants when the fruits are developing would result in lopsided growth, uneven development and ripening of fruits. This operation becomes more important in ratoon crops, as the base of the plant shifts-up, crop after crop. High-density planting would minimize the necessity of this operation, as the plants prop each other preventing lodging.

Water Management

Tripura Queen pineapple is grown mostly as a rain-fed crop in Tripura because the region receives ample rainfall. However, irrigation is necessary during dry spell, especially November to March at 20- 25 days interval ensure good crop. Off season production is not possible without 4-6 irrigations in dry hot months, since stomata of pineapple leaves never open during sunlight due to xerophytes nature of plants. The roots are very sensitive to water logging therefore, tilla (small hillocks) land is most suitable due to good drainage. Tripura is one of the leading pineapples growing state in the country. There are two varieties of pineapples available in the State, Kew variety which is the processing variety and the Queen variety, the table variety normally used for fresh consumption. The Pineapple season in Tripura ranges from the month of June to December with July-August as the peak months.

Flower regulation using growth regulators

Normally, pineapple flowers 10-12 months after planting and fruits are ready for harvest after 5-6 months of flowering. Besides time of planting, flower induction practices with certain chemicals or growth regulators influence the season of harvest. 25 ppm ethephon to mature plants with 39-42 leaves after 7-8 planting or Ethrel @ 0.025 ml/l with 50l cow dung solution can be applied at centre portion. Ethrel @ 0.025 ml/l + 2 % urea + 0.04 % Sodium carbonate induce more than 90 % flowering, as urea helps in better absorption and sodium carbonate increases release of ethylene. Therefore for 1000 plants, 50 liters of solution has to be prepared by dissolving 1.25 ml of ethrel + 1 kg of urea + 20 g of sodium carbonate. Around 50 ml of this solution has to be poured in the crown. Similarly, 95 per cent of flowering within 56 days could be induced by applying 50ml of solution containing calcium carbide (1%) and urea (2%) at the age of 12 months. In ratoon crop of Giant Kew, calcium carbide (1%) and urea (2%) applied in the whorl of plants in September- October induced flowering within one month and the fruits were ready for harvesting in first week of March to end of May. The planting and flower induction schedule can be planned as follows so as to accomplish staggered harvest and unbroken market availability throughout the season ie. From october to May. Although, growth regulators or chemicals can induce flowering at any stage of the plant growth, forcing the plants to produce flowers at an early stage reduces fruit size. Therefore, when plants of optimum size (35- 40 fully grown and active leaves) are induced to flower, better fruit size is obtained without any adverse effects on the ratoon crop.

Harvesting and Handling of Pineapple

The period between planting and harvesting of pineapple is usually two to two and half years. The stage of maturity at harvest is dependent on the required storage or shelf-life and the method of transportation to the export markets. The level of yellow coloration of the "eyes" of the fruit judges maturity. Various colour stages are categorized as follows (Bhattacharjya Etal.1992)

i. CS1: all eyes green, no traces of yellow;

ii. CS2: 5 to 20% of the eyes yellow;

iii. CS3: 20 to 40% of the eyes yellow;

iv. CS4: 40 to 80% of the eyes yellow;

v. CS5: 90% of eyes yellow, 5 to 20% reddish brown;

vi. CS6: 20 to 100% of eyes reddish brown

Fruits are mainly harvested during July-August. However, a small crop is harvested during December to March also. By regulating the crop, harvesting is possible almost 8 month a year. Sugar content should be assessed in the field prior to harvesting to ensure adequate sugar development. A minimum of 10% is generally required although this may vary with the market. Sugar content is not always related to the colour stage as agronomic and production factors will affect sugar development. For the export market where sea-shipment for seven to fourteen days is used, fruits should be harvested at CS1, where the fruits show no yellow colour development on the eyes (ensuring that checks have been made on the sugar content). For air-freighted shipments, although generally cost prohibitive, harvesting can be carried out at CS2 to 3. Those harvested at more advanced stages are more susceptible to mechanical damage and over-ripeness. Fruit maturity can also be assessed on random samples by determination of the flesh condition. This is carried out by slicing the fruit horizontally at the point of largest diameter; in fruit for sea-shipment export; the fruit should show limited development of translucent areas. Where more than half of the area is translucent, the fruit is considered beyond optimum maturity. Pineapples harvested by hand are snapped from the stalk using a downward motion. The fruit should be placed in field crates and while in the field, left in shaded conditions. Collection in the field and field to pack house transport using sacks or bags will cause mechanical damage and increase the level of rejection. On arrival at the packing facility, the stems and the crowns should be trimmed to 2 cm (0.5") and 10 cm (4") respectively. Out grading should be made of all fruits which are undersize, oversize, over-ripe, under-ripe (depending on the market requirements), damaged, bruised or show fungal or insect damage.

Yield

According to records of the Agriculture department, Tripura's pineapples production estimated to be around 1, 26,994 tonnes during 2017-18. The area under cultivation is estimated every year across 8,728 hectare orchards in different districts including Sonamura in Sepahijala district, Kamalpur in North Tripura and Ambassa in Dhalai district, while the productivity is around 14.55 MT/ha. While some of the orchards are run by the government, most of them are privately- owned. Over 4,000 pineapple growers are directly connected to the cultivation of the fruit. Two cultivars of pineapples are largely grown in the State, Kew variety which is the processing variety and the Queen variety, the table variety normally used for fresh consumption. The pineapple season in Tripura ranges from the month of June to December with July to August as the peak months. Economic life of a pineapple plantation is expected to be around 4 years considering one main crop followed by two ratoons as the one crop cycle. On completion of one crop cycle the plantation is uprooted and the land is cleared and prepared for replanting adopting various soil fertility as well as soil amendment practices.

Marketing and Export

There is always a very good demand of Indian pineapples in the internal markets. It is in high demand from the processing industry as well. Indian pineapple is exported to Nepal, U.K., Spain, Italy and U.A.E. The main products of export are canned slices, tit bits, juice etc.

Pineapple based industry in Tripura

Tripura is one of the largest pineapple growing states in the country and there are more than 100 commercial scale pineapple plantations spread throughout the state. Tripura is turning into a commercial hub with large market in Japan, European countries and even in United States. Experiments on high density planting, staggered planting and use of flower inducing hormones are being conducted to increase the productivity of pineapples at the Horticultural Research Centre at Nagicherra, in West Tripura district. Preliminary results indicate the possibility of advancing the availability of pineapples by about two months before the normal season. Pineapple growers in the state have benefited from these results in terms of yield and increased availability period of fruit during the year. It has also helped in engaging the workers throughout the year, thus creating employment opportunity. This method was first introduced only in 25 ha of farm land in the year 2005 but it has now extended to 200 ha area.

Tripura State Horticulture Corporation is gearing up to sign a MoU with Dabur h/t. Ltd, Dharmapal Satpal Group of Industries and North Eastern Regional Marketing Company (NERAMAC) to sell raw pineapple to the tune of 71, 000 t/year. NERAMAC had set up a fruit juice concentration plant, in the year 1988, in Nalkata with an initial capital investment of Rs.30 Million. The Tripura Small Scale Industries (TSSI) and NERAMAC are utilizing Tripura pineapples for making juice, jelly and concentrates. The biggest fruit processing units in Bangladesh, PRAN also utilizes pineapples of Tripura for making vermin products. For the last few years NEMMAC- the Central Agricultural Marketing Agency and Tripura Small Industries Corporation have come forward to help the grower by establishing mini-fruit processing units. But the capacity of these units is inadequate. Meanwhile, a Primary Agricultural Credit Society (PACS) has been set up to look after the marketing aspect of pineapples grown in Durchawi. Last year, the PACS arranged for the export of 1,200 million pineapples to Bangladesh.

Pineapple growers in Tripura are gearing up to meet the Italian demand for canned and bottled pineapple products. The arrival of Pijush Agro Tech, a fruit processing unit in Ushabazar area in Agartala, provides the growers of pineapple with a new ray of hope. Recently a tribal women group from Tripura's pineapple growing tracts is to put to use the technical back-up and training received from the South India Textile Research Association (SITRA) for commercial exploitation of pineapple leaf fiber (PALF).

AEZ of Pineapple

The Ministry of Commerce and Industry has recently sanctioned the Agri Export Zone Scheme for the entire North East Region at Tripura. Under this scheme, enhanced international market access would be provided to the farmers besides necessary infrastructure, flow of credit transport assistance and other facilities for promoting agricultural export through pineapple cultivation. Apart from that the new export-import Policy of the government has given a major thrust to agricultural export by removing export restrictions like registration requirement, minimum export price and exports through the state agencies. All these initiatives are expected to boost the trade potential of pineapples grown in Tripura and other parts of the region. In view of Tripura's inherent strength with regard to organic cultivation of this crop and also in view of the price competitiveness of pineapple products in the international market it was decided to set up an Agri Export zone for pineapple. The zone will cover districts of Kumarghat, Melaghar and Matabari. This project involves a total investment of around Rs. 16 crores, out of which Rs. 8.11 crores will flow from the Central Government agencies, Rs. 1,16 crores from State Government agencies and the remaining by the private sector.

Future Prospect

Naturally, many of the pineapple gardens were replaced with rubber plantations (Suchiradipta and Sarvanan, 2018). The ones which couldn't be replaced still produced plenty to surpass the demand of the state, which reduced the prices (Market price of pineapple is Rs. 43/kg while retail price is Rs. 49 – 55/kg). In spite of that, with the current prices of pineapple in the state, the unit cost of investment is Rs 40,936 per 0.4 acre (1 kani), which is affordable for a small farmer. The net income per unit is around Rs 64,664 with 147% return on investment. Considering producers only get 30-40 per cent of the consumer's money, better connectivity with consumers and reducing market intermediaries will greatly increase the economic benefits of pineapple cultivation in Tripura (MART, 2011). Lack of processing facilities or export initiatives resulted in wastage of huge quantity of pineapples in the state every year. With fall in rubber prices in the international markets, there has been a new found interest in pineapple production, processing, and export in the state. Presently, 8,848 ha area is under pineapple production with 128,971 tonnes yield (14.43 tonnes/ ha is the average productivity). But so far, inadequate processing facilities in the state, so, rise in production without proper market or cold storage infrastructure will result in distress sale of the surplus (currently the wastage is up to 30 %), resulting in a more significant dip in interest of the growers to take up pineapple production. In the recent times, climate change is also adding to the woes of the pineapple farmers (Business Standard, 2016). Tripura has been notified as an Agri-Export Zone (AEZ) for pineapple but except recent exports of few consignments to Dubai, the export to Bangladesh is also at standstill aggravating the market glut. Pineapple exports to Bangladesh have been stopped for past few years which have hit the pineapple growers. Agricultural and Processed Food Products Export Development Authority (APEDA) has exported two consignments (1.05 tonnes each) of Kew variety of pineapple to Dubai in 2017 and based on the positive feedback and high demand, agreements for more are also underway (www.Silchar.com,2018)

Table 1: Quality parameters of queen pineapple in NER as per ICAR research complex for NEH region

State	Variety	Wt. of fruit (kg)	TSS (%)	Acidity (%)	Total sugar (%)	Vitamin-C (mg/100g)
Manipur	Queen	0.6-1.0	18-20	0.5-0.6	-	-
Meghalaya	Queen	1.0-1.5	18-20	0.4-0.5	14-16	-
Nagaland	Queen	1.2-1.5	18-22	0.3-0.4	14-16	50-54
Tripura	Queen	0.8-1.3	13-17.2	0.6-0.8	13-14	28-30

Table 2: Suitable climatic conditions required for quality pineapple production and that available in Tripura

Categories	**Ideal requirement for pineapple cultivation**	**Climatic condition of Tripura**
Soil pH	Slightly acidic soil with pH range of 5.5 to 6.0 is considered optimum for pineapple cultivation	The pH values of Tripura soils vary from 4.05 to 6.05 and in more than 90 percent of soils the pH value is bellow 5.6.
Temperature	The fruits grows well as long as temperature ranges from 15.5 to 32.5^0 C	The maximum and minimum temperature during winter are 27^{0}C and 13^{0}C and during summer are 35^{0}C and 24^{0}C respectively
Rain fall	Optimum rainfall required for pineapple is 1500 mm per year all through	The amount of average total annual rainfall in the state is 2169.4mm.
Soil	Pineapple grows in almost any type of soil, provided it is free-draining	The fertile soil of Tripura and well drained topography of tilla (small hillocks) land with moderate steep slopes, gentle to moderate slope and rolling topography provide perfect cultivation condition.

Table 3: Effect of plant spacing on yield of pineapple

Plant population per hectare	**Distance plant to plant within a row (cm)**	**Distance row to row (cm)**	**Distance trench to trench (cm)**	**Yield (t/ha)**
43,500	30	60	90	45.2
53,300	25	60	90	51.6
63,700	22.5	60 or 45	75 or 90	61.0

(Source: a technology bulletin for Tripura, ICAR Study, 2005)

Table 4: Effect of time of planting, flower induction practices with certain chemicals or growth regulators on season of harvest

Month of Planting (I year)	**Month of application of growth regulators in the II year**	**Months taken for flower induction**	**Harvesting month**
April	March-April	1-2	October-November
May	April-May	1-2	November-December
August	July-August	1-2	February-March
September	August-September	1-2	March-April
October	September-October	1-2	April-May
November	October-November	1-2	May-June

Table 5: District wise Area & Production of Pineapple during the year 2017-18

Pineapple	South District	Gomati District	Sepahijala District	West District	Khowai District	Dhalai District	Unakoti District	North District	State Total
Area (ha)	935	895	750	750	610	2503	884	1401	8728
Production (mt)	13093	12883	10995	10973	8857	36324	12783	21087	1,26,994
Productivity (mt / ha)									14.55

(*Source: Department of Agriculture, Tripura, 2019)*

Table 6: List shows the leading catchment areas of markets of Pineapple in Tripura in the three major districts.

District	**Principle catchment Areas**
North Tripura	Dharmanagar, Unakoti, Fatikrai, Kumarghat, Vanghmuri, Phuldurgsai, Sakhan
Dhalai Tripura	Kamalpur, Halhari, Salema, Kanchanpur, Bahudurpura, Sakhn, Rabiraipara
West Tripura	Khowal, Sidhai, Kalyanpur, Ranibazaar, Jambal, Bisalgarh, Barjula, Sonamura, Kathalia, Teliamura
Unakotu	Darchoi, Jamtailbari, Deocherra

References

Bhattacharya, R. K. and Bhattacharyya, A. P. 1992. Crop production and harvest index of Kew pineapple as affected by foliar application of micronutrients. Acta Horticulturae (ISHS) 296:161-164

Business Standard, 2016. Pineapple growers are in 'double trouble' in Tripura. http://www .business- standard.com/ article/news-ani/pineapple-growers-in-double- trouble-in-tripura-116070900506_1. html (Updated on July 9, 2016; Accessed on May 22,2020).

MART. 2011. Livelihood Based Agri Business and Market Studies for North East Rural Livelihood Project: Final report, Tripura. MART, Noida – 201301, India.

National Horticulture Board. 2017. Area and Production of Horticulture Crops - All India 2015-16 to 2016-17(Final).http://nhb.gov.in/statistics/State_Level/2016-17 (Final).pdf.

Tripura exports pineapples to Dubai. 2018. http://silchar.com/tripura-exports-pineapples-to-Dubai/ (Updated on January 4, 2018; Accessed on May 21, 2020).

22

Organic Farming A Key to Sustainable Environment

P. Sahu, S.P. Monalisa and B. Tripathy

Introduction

Agriculture is an important sector in Indian economy, 58 per cent of India's population depends on agriculture for their livelihood. Agriculture is main victim of climate change, it can disrupt availability, accessibility and quality of food. Moreover, increase in temperature, reduction in ground water availability, flooding and salinity are the major limiting factors in sustaining and increasing vegetable productivity. As a result of this crop physiology and production both are affected. Organic farming can help to tackle climate change by reducing greenhouse gas emissions and amount of nitrogen fertilizer applied to farming land. In India we have advantage that the agricultural techniques we practiced before green revolution in our country is eco-friendly and today also it is somehow it is in memories of our older and younger farmer generation. Principle of sustainable farming is the base of organic farming. According to OECD (Organization for Economic Cooperation and Development) "Sustainable agro-ecosystem - agricultural and food systems are economically viable, meet society's need for food assurance, while they retain and enhance natural resources and environmental quality for future generations" it is very clear from the definition that organic farming mostly emphasizes the environmental aspect and it includes:-

Soil health Enhancement and Maintenance

Soil is most important primary natural resource for any agricultural activity. Organic farming promotes the use of compost, vermin compost and other organic sources of nutrient instead of using synthetic fertilizer. Erosion and conventional farming techniques which is highly dependent on synthetic/ chemical inputs have caused the earth to loose 1/3rd of its cultivable land. Organic farming system is based on the management of soil organic matter, which in terns helps to maintain the physical, chemical and biological properties of soil

(Biswas *et al.*, 2014). Soil microorganism population positively influences soil quality and plays a vital role when assessing the soil productivity. Many researchers reported incorporation of organic inputs in soil increases 20-30% microbial biomass and 30-100% microbial activity. According to Tahat *et al.* (2020), healthy soil acts as a dynamic living system that delivers multiple eco system services, which includes sustain water quality and plant productivity, controlling soil nutrient recycling deposition and also helps in removing greenhouse gases from atmosphere.

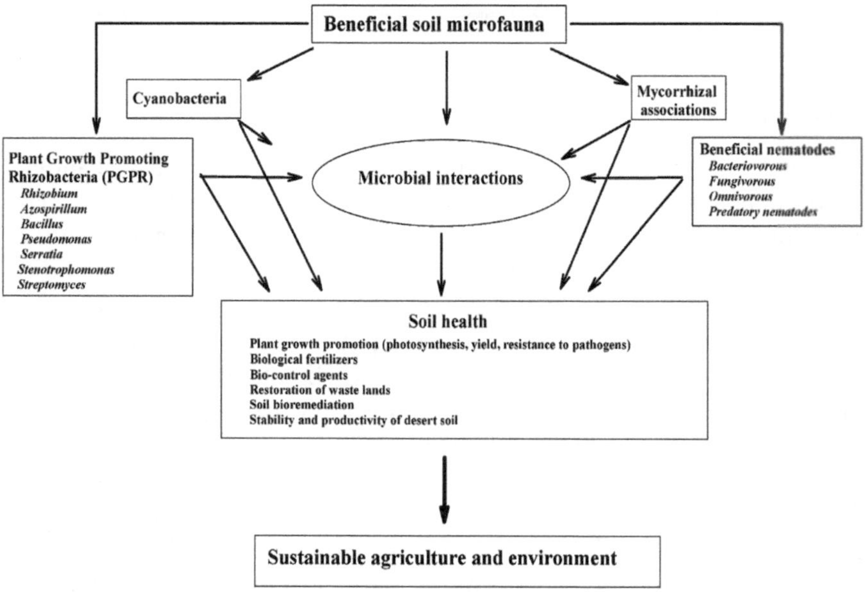

Fig. 1: A conceptual theme demonstrating the role of beneficial soil microbes and their interactions for the development of sustainable agriculture and environment
(*Source*: Tahat *et al.*, 2020)

Recycling or recirculation of nutrient and other sources

In conventional farming use of soluble form of nutrient i.e., synthetic/chemical fertilizer is authoritative, over use of mineral fertilizer may result good harvest in initial years but results depletion in soil nutrients, destroy biotic flora of soil environment, land degradation, decrease N use efficiency and eutrophication in crop in long run (Pathak *et al.*, 2016). On the other hand natural ecosystem process depends on natural biological processes such as nitrogen fixation and soil organic matter mineralization. Manure nutrient converted to more stable form which prevents loss of nutrient through surface runoff and other process. Use of organic source as fertilizer used to release nutrients slowly to the plant and improve nutrient retention in soil. According to Hait and Tare, 2012, composting and vermin composting are recycling organic waste in agricultural field can be a sustainable alternative for replacing synthetic fertilizer, also listed multiple benefits like structure and minimizes erosion of soil, improvement in soil nutrient profile, wastage reduction from dumpsite results in reduced environmental pollution.

Water conservation and water health

In many parts of the world the quality of surface and sub-surface water has been jeopardized by intensive land management practices and the consequent imbalance of C, N and water cycling in soil (Doran and Zeiss, 2000). Indiscriminate use of synthetic fertilizer and pesticide in agricultural sector is main cause of water pollution. In a study it was reported 35- 65% less nitrates loss though organic farming when compared to conventional agriculture.

Reduction in greenhouse gas emission and improvement in air quality

Agriculture plays both source and sinks for production of GHGs (Green House Gases), carbon emission through direct use of fossil fuel in agricultural operation and indirectly through intensive use of fertilizer and continuous cultivation of soil resulting in loss of soil organic carbon/organic matter in conventional agriculture. The present threat of global climate change and ozone layer depletion, through elevated levels of greenhouse gases and altered hydrological cycles, creates a urgent necessitates for better understanding of the influence of land management on soil processes (Bengtsson, 1998). Organic farming facilitates biomass incorporation in soil through crop rotation, cover cropping, mulching, recycling of nutrient (composting, vermicomposting etc.), help in build up soil organic matter in above ground/surface soil. So consider as one of the solution to mitigate climate change because it is believed to emit much lower amount of GHGs and also efficiently sequesters carbon in soil (Panwar *et al.*, 2010). Also organic land traps more carbon in to humus as a result reduces atmospheric CO^2 and contribute towards a stabilize ecosystem.

Maintaining biodiversity and protection to genetic resources

Intensive agriculture includes monoculture and replacement of traditional varieties with application of heavy synthetic fertilizer and pesticide is now creating greatest threat to world wide biodiversity and food nutrition security (Kavitha and Chandran, 2017). Organic farmer are conservator and user of biodiversity in different levels. Traditional and wild cultivars are preferable option while considering gene level, as these breeds shows higher degree of resistance to disease and also resilience to changing climate.

Efficient use of energy and focusing on renewable resources

Organic farming system make farms and people more resilient to climate change because of low risk farming strategy with reduced input cost, so lower down the risk of partial or total failure due to climate change (Patle *et al.*, 2014). The renewable energy includes/sustainable includes hydropower, tidal power, geothermal energy, solar energy, energy obtained from biomass etc, is obtained from a non-exhaustible resource, so efficient use of this energy can solve the present need without compromising the need of coming generation. Therefore in practices of energy conservation organic farmers can use of renewable energy to reduce dependency on external sources and become self-sufficient. Further, it helps in saving capital as well as combats the harmful effect of global warming.

Optimization of life for all organisms including humans

Organic farming facilitates better and efficient use of natural resources as a result the system provide food and shelter to existing species in that habitat, also restriction on use of harmful chemical attracts new species (wild flora and fauna), facilitates growth of beneficial microorganism in soil. Areas included under organic farming are more diverse i.e., 85% greater number of plant species, a third more bats, 17% more spider and 5% more bird species as compared to conventional farming areas (Hole *et al.*,2005).

Conclusion

In the era of deforestation and urbanization / industrialization many changes in climate is observed over a long period, climate change is a slow process and can be reversible by adapting long term climate resilient agricultural techniques. Organic farming can open a path for providing nutritionally rich food without adversely affecting environment. Now- a- days, health is more concerning factor for world, and organic food products are capturing the market, as people are aware about the harmful effect of pesticide, insecticide and other synthetic chemicals in their serving plate. As a consumer we can select organic products according to our purchasing capacity and can contribute towards selecting a less polluting option of agriculture as organic farming.

References

Bengtsson J.1998. Which species? What kind of diversity? Which ecosystem function? Some problems in studies of relations between biodiversity and ecosystem function. Applied Soil Ecology, 10, 191–199.

Biswas S, Ali MdN, Goswami R and Chakraborty S. 2014. Soil health sustainability and organic: a review, Journal of Food Agriculture and Environment, 12(3&4); 237-243.

Doran J and Zeiss MR.2000. Soil health and sustainability: managing the biotic component of soil quality, Applied Soil Ecology, 15; 3–11

Hait S and Tare V. 2012. Transformation and availability of nutrients and heavy metals during integrated composting-vermicomposting of sewage sludge, Ecotoxicology and Environmental Safety,79;214-224.

Hole DG, Perkins AJ, Wilson JD, Alexander IH, Grice PV and Evans AD.2005. Does organic farming benefit biodiversity?, Biological Conservation, 122;113-130.

Kavitha V and Chandran K.2017. Organic farming in conserving biodiversity in India-a review, Agricultural Reviews, 38(04). doi:10.18805/ag.R-1709.

Panwar RP, Singh NR, Ramana AB, Yadav S, Shrivastava SK and Subbarao R.2010. Status of organic farming in India, Current Science, 98;1090-1194.

Pathak H, Jain N, Bhatia A, Kumar A and Chatterjee D.2016. Improved nitrogen management: a key to climate change adaption and mitigation. Ind J Fertil, 12(11);151-162.

Patle GT, Badyopadhyaya KK and Kumar M. 2014. An overview of organic agriculture: a potential strategy for climate change mitigation, Journal of Applied Natural Science, 6(2); 872-879.

Tahat MM, Alananbeh KM, Othman YA and Leskovar DI. 2020. Soil health and sustainable agriculture, Sustainability, 12(4859); 1-26. doi:10.3390/su12124859.

23

Influence of Pink Pigmented Facultative Methylotrophs (PPFMS) and Plant Growth Promoters on Availability of Soil Macro Nutrients Status on Semi Dry Rice

B.B. Priya and G. Balasubramanian

Introduction

Rice, being an important crop in India, there is a part to focus on improving rice production and productivity. A number of biotic and abiotic factors mostly referred as production constraints that affect the rice yield adversely and subsequently, needs an extraordinary consideration to the developing demand of continuously emerging population of India (Kumar and Jha, 2000). Abiotic stress has been estimated that more than half of the yield potential of major crops is mainly affected due to unfavourable growing environments such as drought (Cortina and Culianez-Macia, 2005). The current study was aimed to carry out to assess the impact of Pink pigmented facultative methylotrophs (PPFM) and plant growth promoters on mitigating drought stress effects in semi dry rice.

Materials and Methods

The field experiments were conducted in sandy clay loam soil of farmers field, Therkutheru, Melur block, Madurai during 2017-18 with the short duration variety ADT-45. The experiment was conducted on the basis of factorial randomized block design with three replications. The experiment was conducted with the following treatments viz., Factor 1: Soil Application S1: 100% Recommended Dose of Fertilizer (75:25:37.5 kg ha-1N, P and K), S2: 100 % RDF+ *Azospirillum* (soil application @ 2 kg ha^{-1}), S3: 100 % RDF+ Azophos (soil application @2 kgha^{-1}) and S4:100%RDF+Azotobacter (soil application @ 2 kg ha^{-1}), Factor 2: Foliar application F1: PPFM (Foliar application of 500 ml ha^{-1}), F2: Brassinolide (1 ppm), F3: Salicylic acid (100ppm) and F4:Control (No spray).

Results and Discussion

Soil available nitrogen

The pooled field experimental data of available macro nutrient status showed a significant effect at harvest stage by the combined application of soil and foliar application (Table 1,2 and3). From the field experiments, it was recorded that the application of RDF + Azophos @ 2 kgha-1 (S3) through soil recorded the mean values of 246 to 243, 237 to 232 and 213 to 202 kg per hectare at 30 DAS, 60 DAS and harvest stage, respectively. Among the different treatments, foliar application of PPFM @ 500 ml ha-1(F1) the available nitrogen mean values recorded 250 to 206, 234 to 198 and 194 to 180 kg ha^{-1} at 30 DAS, 60DAS and harvest stage respectively. Further the data appeared a concomitant increase with respect to available nitrogen by the combined application of soil and foliar application registered in the treatment (S3F1) soil application of RDF + Azophos @ 2 kg ha^{-1} along with foliar application of PPFM @ 500 ml ha^{-1} (S3F1) in available nitrogen value of240, 217and 168kg ha^{-1} at 30 DAS, 60 DAS and during harvest, correspondingly, which was followed by the treatment application of RDF + Azophos @ 2 kg ha-1through soil along with foliar application of Brassinolide @ 1 ppm (S3F2) for all growth stages.The highest available nitrogen value of 246, 237 and 213 kg ha^{-1} were recorded in control at 30 DAS, 60 DAS and during harvest, respectively. Facultative methyl bacteria can fix nitrogen under chemolithtrophic conditions in the rhizosphere of rice paddy soils and provide the rice plants with nutrients (Oyaizu-masuchi and Komagata,1988).

Table 1: Effect of PPFM and growth promoters on available soil nitrogen (kg ha^{-1}) in semi - dry rice (pooled data of two years)

Soil applicati on	**Available Soil Nitrogen (kg ha-1)**														
	30 DAS					**60 DAS**					**At harvest**				
	Foliar application					**Foliar application**					**Foliar application**				
	F1	F2	F3	F4	Mean	F1	F2	F3	F4	Mean	F1	F2	F3	F4	Mean
S1	260	262	264	201	246	249	253	255	191	237	218	223	228	183	213
S2	247	249	251	208	239	230	232	236	199	224	187	192	197	175	188
S3	240	243	245	212	235	217	223	226	205	217	168	176	181	178	176
S4	254	256	258	206	243	241	243	246	198	232	204	209	212	184	202
Mean	250	253	254	206		234	238	241	198		194	200	204	180	
	S	F	Sx F			S	F	Sx F			S	F	Sx F		
SEd	2.19	2.19	4.38			2.25	2.25	4.50			4.15	2.7	4.15		
CD (P=0.05)	4.47	4.47	8.94			4.59	4.59	9.10			8.48	4.24	8.48		

Table 2: Effect of PPFM and growth promoters on available soil phosphorus (kg ha^{-1}) in semi - dry rice (Pooled data of two years)

Soil application	**Soilavailablephosphorus(kgha-1)**														
	30 DAS					**60 DAS**					**At harvest**				
	Foliar application					**Foliar application**					**Foliar application**				
	F1	**F2**	**F3**	**F4**	**Mean**	**F1**	**F2**	**F3**	**F4**	**Mean**	**F1**	**F2**	**F3**	**F4**	**Mean**
S1	37.3	37.5	38.2	21.3	33.5	32.3	32.1	32.3	16.7	28.3	28.6	29.0	29.3	16.3	25.7
S2	33.3	34.1	34.7	22.7	31.1	28.0	28.6	29.7	17.8	26.0	24.8	25.9	25.7	16.6	22.8
S3	32.3	32.7	32.5	23.3	30.2	26.3	27.7	27.9	17.6	24.8	23.9	24.2	24.6	16.8	21.9
S4	35.0	36.3	36.7	22.2	32.5	30.1	31.3	31.5	17.1	27.3	26.3	27.3	27.7	16.5	24.3
Mean	34.3	35.1	35.5	22.3		29.2	29.8	30.3	17,3		25.6	26.3	26.5	16.4	
	S	F	S x F			S	F	S x F			S	F	S x F		
SEd	0.42	0.42	0.84			0.28	0.28	0.56			0.29	0.29	0.59		
CD (P=0.05)	0.85	0.85	1.71			0.57	0.57	1.14			0.69	0.69	1.31		

Table 3: Effect of PPFM and growth promoters on available soil potassium (kgha^{-1}) in semi- dry rice (Pooled data of two years)

Soil application	Available Soil Potassium (kg ha^{-1})														
	30 DAS					60 DAS					At harvest				
	Foliar application					Foliar application					Foliar application				
	F1	F2	F3	F4	Mean	F1	F2	F3	F4	Mean	F1	F2	F3	F4	Mean
S1	351	354	357	304	342	329	332	336	286	321	311	314	318	268	303
S2	334	338	341	310	331	304	310	313	290	304	287	292	295	273	287
S3	323	329	333	318	326	289	295	301	297	295	271	278	283	280	278
S4	343	346	350	307	336	317	322	325	287	313	299	304	307	270	295
Mean	338	342	345	310		309	315	319	290		292	297	301	273	
	S	F	S x F			S	F	S x F			S	F	S x F		
SEd	2.98	2.98	5.97			2.92	2.92	5.84			3.10	3.19	6.38		
CD (P=0.05)	6.10	6.10	122			5.96	5.96	1.19			6.51	6.51	1.30		

Soil available phosphorus

Among the different sources, soil application of RDF + Azophos @ 2 kg ha^{-1}(S3) recorded minimum values of 30.2, 24.8 and 21.9 kg ha^{-1} at 30 DAS, 60 DAS and during harvest, correspondingly. The experimental data revealed that PPFM and growth promoters application available phosphorus mean values ranged from 34.3, 29.2 and 25.6 kg ha^{-1} at 30 DAS, 60 DAS and harvest stage respectively and the higher value of 33.5, 28.3 and 25.7 kg ha^{-1} at 30 DAS, 60 DAS and during harvest, respectively was recorded in control plot. Further the data showed a concomitant increase of available phosphorus by combining the application of soil and foliar application which recorded higher available phosphorus of 32.3, 26.3 and 23.9 kg ha^{-1} (S3F1) soil application of RDF + Azophos @ 2 kg ha^{-1} along with foliar application of PPFM @ 500 ml ha^{-1} (S3F1). It was more effective than individual application of biofertilizers as suggested by Kucey (1983) directly solublize the rock phosphate bring the phosphate into easily available to plants.

Soil available potassium

Among the various treatments, soil application of RDF+ Azophos @ 2 kg ha-1 (S3) recorded the higher mean values of soil available potassium ranged from 326,295, and 278 kg ha-1 at 30 DAS, 60 DAS and harvest stage, respectively. This was followed by the treatment S2 that received soil application RDF+ Azospirillum @ 2 kg ha^{-1}. Application of PPFM@500 ml ha^{-1} (F1) have recorded maximum soil available potassium mean values of 338,309 and 292 kg ha-1at 30 DAS, 60 DAS and harvest stage, respectively. This was followed by the mean values of 342,315 and 297 kg ha-1 at 30 DAS, 60 DAS and harvest stage, respectively in the treatment that received foliar application of Brassinolide @ 1 ppm (F2) and the lowest soil available potassium was recorded in control. The results of the experiment pertaining to the conjoint effect of soil and foliar application on soil available potassium was found to be statistically significant. The effect was found to be in the highest in the treatment that received application of RDF+Azophos @ 2kg ha^{-1} through soil along with foliar application of PPFM @ 500 ml ha-1 (S3F1) which recorded maximum soil available potassium values of 323, 289 and 271 kg ha^{-1} at 30 DAS, 60 DAS and during harvest, correspondingly and this was followed by the treatment, soil application of RDF+ Azophos @ 2 kg ha^{-1}along with foliar application of Brassinolide @ 1ppm (S3F2)329,295 and 278 kg per hectare at active tillering, panicle initiation and during harvest respectively. Similar findings was also reported by (Nagendra, 2019) which indicates that the microorganisms in the soil can solublize unavailable forms of potassium bearing minerals such as micas, illite and orthoclase, by excreting organic acids which either directly dissolves rock potassium or chelate silicon ions to bring the potassium in to solution (Bennette, 1998).

Thus, agricultural productivity is gradually decling year by year due to many abiotic stresses especially drought. Of the various management practices, increased availability of mitigation through plant growth promoters and bio-products like PPFM are promising to enhance the productivity and yield of rice crop. Therefore, these results have practical field application to protect the plant under drought especially PPFM. Further studies would be required to confirm the effect of PPFM through molecular studies.

References

Bennette, R. 1998. "The Meaning of Dress: Nationalism, Feminism, and Fashion In: "Early Twentieth-Century Ireland." Proceedings of the Harvard Celtic Colloquium.

Cortina, C. and Culianez-Macia, F.A. 2005. Tomato abiotic stress enhanced tolerance by trehalose biosynthesis. Plant Sci. 169(1): 75-82.

Kucey, R. 1983. "Phosphate-solubilizing bacteria and fungi in various cultivated and virgin Alberta soils." Canadian Journal of Soil Science 63 (4):671-678.

Nagendra, D. 2019. "Influence of pink pigmented facultative methylotrophic (PPFM) bacteria and pseudomonas flourescens on blackgram (Vigna mungo L.) under drought conditions." Acharya NG Ranga Agricultural University, Guntur.

Oyaizu-masuchi, Y., and K. Komagata. 1988. "Isolation of free-living nitrogen-fixing bacteria from the rhizosphere of rice." The Journal of General and Applied Microbiology 34(2):127-164.

Qayyum, M. A. and Malik, M. D. 1988. Farm production losses in salt affected soils. Proc.1st Nat.Congr. Soil Sci. Lahore, Oct., 356- 364.

24

Regenerative Agriculture – A Futuristic Approach to Solve Emerging World Food Crisis

N. K. Dhal and J. M. L. Gulati

Introduction

Agricultural production system achieved a miracle by the way of green revolution which compactualized the potential of photoperiod and the thermo insensitive input responsive crop culture. The productive trends of food crops continued with increased fertilizer consumption, irrigation potential and improved agro-techniques. However science based intensive agriculture with intensive irrigation due to its inappropriate, inpropotional and ambitious approaches put the agriculture productivity on back foot. Because of these problems the arable land is facing the problem to sustain itself in terms of its capacity to produce desirable output. The scenario is further threatened with increase in climate change impact on various natural resources. The projected food crisis is on the card.

The Challenging Issues

1. Food crisis and food security

The world is going to face food crisis in near future. The 2022 projection showed that feeding a world population of 9.1 billion people by 2050 would be required to raise the overall food production by some 70%. Production in developing countries and territories would need to be almost double. Many of the African countries are presently experiencing food crisis. As per the current status, India has just enough food grain to feed its people. We are certainly not out woods when it comes to the consequences of global hunger of our life time. The global multidimensional poverty index (MPI) 2022 indicated that 111 countries in developing regions are home to 6.1 billion people are under poverty line. Similarly, according to 2022 Global Hunger Index India ranks 107 out of 121 countries with a score of 29.1. India has a level of hunger that is of serious concern when we compare India which ranked 4th among the countries having more number of billionaires (124).

2. High food pricing

Currently, it is the food prices that are the challenge and not availability, according to WFPs Chief Economist, Anil Husain. High food prices have triggered a global crisis that is driving millions more into extreme poverty, magnifying hunger and malnutrition. A number of people, who are experiencing acute food insecurity and will need urgent assistance, which likely climb to 222 million people in 53 countries and territories, according to a FAO – WFP report (2022). But, how long? Time is approaching very first when people belonging to higher income group mostly will have access to high priced food. A sincere effort must be made by the agricultural scientists to develop technology to increase food production to create buffer stock of staple foods in order to check food price and make it available to people of lower income groups in affordable price.

3. Natural resource hazards

It is high time to realize that agricultural practices followed to provide not just agricultural produce but also ecosystem services like carbon sink, water recharge, and evolutionary potential. Commercial agriculture that dominates food and fiber supply chains, incentivizes practices that promote soil erosion, nutrient runoff and harmful algal blooms in fresh water and coastal streams. Thus, the present day agriculture has compartmentalized natural resources and focused on the yield of individual crops.

4. The thought of the day

Increase in food production from the present level to the projected level of 70% or more is not an easy task due to adverse production factors like, climate change, declining in soil nutrient availability, in many regions the agricultural land is at the risk of conversion to suburban and urban development leading to more and more areas becoming unavailable for food production, various types of soil erosions and many more. We must have to deploy nature based solutions to support agricultural production while helping regeneration ecosystem. The nature based solutions involve conserving or rehabilitating natural ecosystem and enhancement or creation of natural processes in modified or artificial ecosystems. We need to act now before the impact of climate change and degradation of natural resources is worsen, leaving more families vulnerable and hungry, and eroding the region's hard earned prosperity. The author feels the best way to counter the basic problems is to adopt "regenerative agriculture" and make the life of every living system on this earth, starting from microorganisms to human beings and animals be sustainable. Regenerative agriculture also can substantially mitigate climate change impact besides providing food for every one.

What is Regenerative Agriculture?

Robert Rodale in 1980 coined the term "regenerative agriculture (RA)" to describe a holistic approach to farming that encourage continuous innovation and improvement of environment, social and economic measures. RA is a conservation and rehabilitation approach to food and farming system that among other benefits, reverse climate change by rebuilding soil organic matter and restoring degraded soil biodiversity - resulting both carbon drawdown and improving the water cycle that helps in achieving sustainable long term productivity to achieve the desired level of crop yield. Thus, the RA is aimed at improving soil productivity and slowing down the process of climate change.

The climate change impact

The mismanagement of resources and anthropogenic factors has led to global warming. Increased carbon dioxide concentration along with emission of other green house gases has caused an increase in global surface temperature by 0.76^{0} C with predicted additional rise of 1.3^{0} C to 1.8^{0} C by 2050 (IPPC, 2007). An altered hydrological cycle is being also observed since recent past due to climate change. A shift in weather pattern especially of rainfall both in terms of intensity and duration has been a regular feature at macro and micro scales. Occurrence of extreme events has led to either flood or drought like situations. Cloud bursts in random localities, is a cause of soil nutrient runoff, landslides in hilly tracts and excess soil erosion. It is because of the fact that the soil is unable to absorb the total rain falling at that particular time. On the other hand limited or scanty rainfall besides inducing soil moisture deficit and drought like situations of different magnitudes also stimulates heat waves which is another climatic hazard being observed in coastal areas in recent past.

The Soil Factor

The Director General, FAO has said that "The multiple roll of soils often go unnoticed. Soils don't have a voice, and few people speak out for them. They are our silent ally in food production." Physical condition of soil and biological fertility are normally overlooked in soil health management by the soil users. Recognizing the importance of soil health the 68th session of UN General Assembly declared the year, 2015 as the 'International year of Soil'. It is being observed that the productivity of soil is gradually decreasing due to imbalance in various physical, chemical, biological and nutritional properties. This has resulted in slow permeability of soil due to high clay content leads to more runoff, soil erosion and nutrient removal from soil; the colloidal oxides of iron and aluminium in soil binds the soil particles under wet condition and dry up to form crusting of surface soil which affect the crop growth starting from seed germination to harvest; formation of sub soil hard pan due to alleviation of clay to the sub soil horizons which is adversely affecting the water infiltration, reduction in water holding capacity, movement of air and nutrition, slowing down of the process of recycling of soil nutrients that are being carried out by the soil microbes. Thus, the practice of commercial agriculture is gradually transferring the cultivated lands in to problematic soil leaving besides the basic problematic soils like, saline, alkaline, acidic, sandy and waterlogged soils which need special reclamation practices. Soil organic matter is a key indicator as far as soil quality is concerned. The sustainable soil-carbon pool supports a healthy ecosystem, production system coupled with higher input use efficiency and sustainability index. Soil – carbon stock is also related to climatic conditions, temperature and rainfall, besides the cropping pattern followed.

Regenerative agriculture – ways and means

Measures to implement Regenerative Agriculture

Agriculture plays a significant role in contributing to climate change by the process of converting atmospheric carbon dioxide into organic carbon and release part of it into the soil through its roots where it is stored. However process can also be quickly by releasing the carbon dioxide back into the atmosphere through farm practices like plowing and tillage, soil preparation following excessive use of the practice of mechanization leading to breaking of soil texture and structure resulting release of CO2 into the atmosphere. Our food systems are also suffering enormous consequences from in addition to rising temperatures that are themselves changing where and how things can be grown. The measures need to be taken up with the objective to improve long term soil productivity level through improved water holding capacity of soil, ground water recharging capacity, check surface runoff, shift to sustainable cropping pattern, improve soil organic carbon level and mitigate climate change through carbon sequestration. Soil organic carbon is the main component of soil organic matter. Plants through the process of photosynthesis absorbs carbon dioxide from the air and – using the sun's energy, water, and nutrients from the soil transforms it into carbon that the plant uses for growth of leaves, stems, roots and fruit/grains. The excess carbon created through this process is transported down the plant and is stored in the surrounding soil, sequestering the carbon in the ground. This carbon in the soil is known as soil organic carbon and it feeds the soil microbes which in turn provide nutrients for the plant.

Soil reclamation: Hard clay pan formation below the surface soil has become a common phenomenon more particularly in rice – rice cultivation system under intensive irrigation which has slowed down the process of soil water infiltration, biological cycle that is being carried out by the soil microorganisms, increased soil nutrient runoff and erosion. There is need of deep ploughing of land up to 45 cm depth using chiseling technology with mole plough at least once in every 3-4 years and growing of deep rooted crops in rotation.

Reduce soil runoff: The problem persists mostly in sloppy lands where there is need to provide contour bonding across the slope. Also, ridge and furrow method of cultivation practice planting kharif vegetable or direct seeding of grain crops.

No – till farming:. Soil productivity depends on the C-N ratio of soil. Intensive tillage has led to soil erosion and an increased release of carbon dioxide into the atmosphere as the carbon pool of soil comes in contact with oxygen. Wet soil tillage leads to formation of compact layer at plough sole zone causing a reduction in aerated environment which in turn affect microbial activities. Thus, an approach of zero or minimum tillage farming is advocated to take care of the soil by preventing soil erosion with use of live soil cover and minimum soil disturbances. This in turn over the years will increase the level of soil organic matter that will be helpful in providing an ideal soil environment for better plants growth and ultimately an increase in food production to achieve the goal of food security.

Cover cropping: Do not keep the soil bare after the main crop is harvested as a bare land releases more of carbon dioxide in to the atmosphere form soil due to conversion of soil carbon in to carbon dioxide in presence of oxygen. Therefore, the soil should be covered with a second crop, preferably a pulse crop, taking advantage of the field moisture and nutrition or for the purpose, slight change in the crop planning need to be made.

Organic farming: Application of fossil fuel based imputes need to be reduced or be stopped. Practice of application of FYM in crop field at the time of final land preparation at the rate of 12 t/h helps in improving soil organic carbon level and crop productivity.

Agroforestry and Agro-horti based farming system: Such farming practices in large bare lands or deforested area will help long term soil cover and better utilization of atmospheric carbon dioxide and one can be benefited through long term income generation from farming.

Crop diversity: Different plants release different carbohydrates (sugars) through their roots, and various microbes feed on these carbons and return all sorts of different nutrients back to the plant and the soil. Therefore growing of sallow and deep rooted crops through the process of crop rotation in their fields, farmers help to create the rich, varied, and nutrient-dense soils that lead to more productive yields.

Use of large grain seed crop: In red soil where surface crust is formed when the wet soil dry up after every rain and to break open the crust for better soil aeration and plant growth one should raise large grain crops.

Futuristic Research Need

Regenerative agriculture practices have received little attention of researchers. Better land management and agricultural practices could enhance the ability of soils to store carbon and help combat global warming. Agricultural practices that disturb the soil, such as tilling, planting mono-crops, removing crop residue, excessive use of fertilizers and pesticides and over-grazing expose the carbon in the soil to oxygen, allowing it to burn off into the atmosphere. Deforestation and the draining of peat lands also cause soils to release carbon. Therefore there is need to modify our researches on crop and soil management with the objective of enhancing crop production to meet the food production challenges. Let the agricultural management practices play a critical role in enhancing soil carbon sequestration by following the integrated approach in holistic manner.

Conclusion

If you take care of your soil, it will take care of you. Thus, regenerative agriculture is a system of farming principles and practices that seeks to boost climate resilience, reduce green house gas emission, protect and restore natural ecosystem, improve soil health to mitigate climate change impact and at the same time increase food production and preserve the very precious agricultural lands. It is the responsibility of the researchers in the fields of agriculture, the environmental scientists and the policy makers to think for the common men of future generations who are looking at you for food and the well to do people for nutritious food. The solutions for both can only be achieved through regenerative agriculture.

Information Sourses

Ram L. Ray, Richard W. Griffin, Ali Fares, A. Elhassan, S. Woldesenbet and Eric Rish (2020) Soil CO2 emission in response to organic amendments, temperature, and rainfall. Scientific Reports. Vol. 10, 5849.

E - sources
https://hdr.undp.org>2022MPI
https://www.global hungerindex.org
FAO. 2022. FAO publications catalogue 2022 – April. Rome. https://doi.org/10.4060/cb9264en
http://www.agritech.tnau.ac.in/agriculture/agri_reosourcemgt_soil_soil contraints.html
https://www.climaterealityproject.org

25

Enhancing Food Security Through Agroforestry for Sustainability A Review

G. Sahoo and A.M. Wani

Introduction

The COVID 19 pandemic is threatening food systems and international food security. According to the World Food Programme (WFP), that says the amount of individuals facing crisis hunger is anticipated to virtually double this year to 265 million. Already, quite 820 million individuals don't get enough food to eat, and another a hundred thirty five million individuals face acute hunger or starvation. Increase that, the economic destabilization caused by COVID-19, and another a hundred thirty million people are in danger of starvation by the top of 2020, says WFP. The quickness with that a health crisis transforms in an exceedingly hunger crisis shows however fragile is our food systems to shocks of any nature.

Over 800 million individuals were undernourished in 2019 – a majority living in low income countries. With the pandemic, vulnerable households' ability to shop for basic food can decline, similarly as their risk to earn a daily income. several countries full of food insecurity – as an example within the Horn of Africa, West Africa, Latin America and parts of Asia – have key agricultural seasons beginning around world Food production is vulnerable by preventive measures taken against Covid-19 as a result of reduced farmers ability to access inputs, labour or credits. A discontinuous food production chain might reduce confidence within the reliable flow of basic food commodities that might rock the markets and near food costs. Thus, there's an at hand risk that the pandemic can have devastating consequences for individuals living in low-income countries, with long- run repercussions on people's diets and livelihoods. Forests, trees and agroforestry give important contributions to Food Security and Nutrition (FSN). All of those contributions are even additional necessary in times of crisis. They contribute to livelihoods and to the diversification of production and sources of financial gain therefore additionally increasing the resilience of households. They supply ecosystem services -water regulation, soil fertility and conservation, pollination, temperature regulation- that support farming systems and contribute to their adaptation to climate change. They're a necessary element of property and resilient food systems, contributively to the four dimensions of food security and nutrition each for the forest-dependent communities and globally.

Agroforestry, a distributed land management system within which trees or shrubs are grown up around or among crops or grassland – will contribute to food security and additional nutrient diets by contributing to improved diverseness and nutrition, higher yields, and additional various livelihoods for farmers. This combination of agriculture and forestry will build a spread of healthy foods obtainable, whereas at a similar time causative to additional property food production systems that also are additional resilient to environmental shocks and will increase the resilience of vulnerable teams in times of crisis. When done properly, the inclusion of trees in agriculture will guarantee continuing access of nutrient foods for the foremost vulnerable populations, similarly as sustained functions of native food markets, price chains and systems. Whereas hoping on one crop tends to guide to food insecurity and singular financial gain bursts, a diversity of crops will give an additional nutrient-rich diet, shut the hunger gap between harvests of staple crops, and make opportunities to earn an additional stable financial gain.

During this crisis, it's important to support vulnerable smallholder farmers and food staff similarly because the entire food supply particularly for vulnerable urban areas. A number of these ways include; putting in place vegetable kitchen gardens and growing a spread of fast growing indigenous vegetables which don't need a lot of preparations and external inputs like cowpeas, amaranths, jute, cassava leaves. Crops like *Moringa oleifera* take six months to grow and the leaves are extremely rich in nutrient content. Growing of short term trees that give food like pigeon peas is also differently which will facilitate cushion in times of crisis; fodder trees that feed stock for nutrition like bush also will guarantee healthy stock and thus the supply of milk for youngsters and fresh mothers; leguminous plant *Sesbania sesban* and *Calliandra* trees area also fast maturing (less than one year) crops for provision of fuel wood and fodder. The farmers but additionally would like future plans tree institution plans as they grow the short term trees for immediate edges to address the crisis. Farmers will invest in apiculture and tree nursery management for income-generating. As this pandemic has been started since begin of 2020, the most concern is food for sustenance and property for future. During this context, agroforestry play a significant role for manufacturing food, fodder and fuel to satisfy the necessity of the individuals within the world. This review deals with however agroforestry fulfill the necessity of individuals additionally to keep up balance of the setting for property.

This major challenge to world food security is that they ought to about double food production over consequent few decades, particularly thanks to rapidly growing demand from the developing world (Food and Agriculture Organization of the international organization (FAO), International Fund for Agricultural Development & World Food Programme, 2015; Godfray *et al.*, 2010; International Assessment of Agricultural Science & Technology for Development (IAASTD), 2009; The Government Office for Science, 2011; The Royal Society, 2009). To realize yield will increase, the employment of chemical inputs, genetic improvement, and mechanization has currently become typical, because of their success within the past (with the vital exception of Africa; IAASTD, 2009; Pretty and Bharucha, 2014; the Govt. office for Science, 2011; The Royal Society, 2009). However, typical agriculture has conjointly been a principal explanation for varied social and environmental issues as well as temperature change, loss of biodiversity and ecosystem integrity, land degradation, water insecurity, and disruption of social systems (Chappell and La Valle, 2009; Godfray *et al.*, 2010; IAASTD, 2009; Maxwell, Fuller, Brooks and Watson, 2016; Pretty and Bharucha, 2014; Pretty, Morison, and Hine, 2003; The Royal Society, 2009). Consequently, there's currently in depth accord that we want to move away from the present, narrow focus on yield, and toward a lot of "multifunctional" agriculture that conjointly respects (and ideally enhances) broader social group and environmental goals, below the rubric property intensification (IAASTD, 2009; Godfray and Garnett, 2014; Pretty, 2008; Pretty and Bharucha, 2014; The Government Office for Science, 2011; The Royal Society, 2009).

Indeed, the continued degradation of soil and ecosystems threaten the property of food production itself, as will world environmental modification (Amundson *et al.*, 2015; Foley *et al.*, 2011; Montgomery, 2007; Ferdinand Joseph La Menthe Morton, 2007; Potts *et al.*, 2010; Stern, 2007; Tan, Lal, and Wiebe, 2005; The Royal Society, 2009; Tscharntke *et al.*, 2012). Here, we tend to first counsel that the sustainable development goals (SDGs; Griggs *et al.*, 2013; United Nation General Assembly, 2015) offer a broad and coherent framework for multifunctional agriculture, since this international agreement already combines food security (SDG2) with environmental, climate, and sustainable goals and so emphasizes the necessity for a multi goal approach. We tend to then describe however an extremely multifunctional different already exists in agroforestry. We tend to shortly explore however typical agriculture tends to dominate as compared to alternatives like agroforestry, with multiple negative consequences. Finally, we tend to define issues in making a lot of acceptable balance of approaches. One issue with a multi goal approach is that each agricultural possibility can have an effect on every individual goal otherwise (and each actor can provide completely different weight to the action of every goal). However, food is prime to human life, and then we tend to assume that multi goal agriculture should 1st be ready to increase yield sufficiently to satisfy the SDG of food security. Once that criterion is glad, the choices that advance alternative SDGs most powerfully (or compromise them least severely) would receive priority. Additionally, every agricultural approach can have an effect on the SDGs otherwise for completely different social groups and different geographies.

Since food security and therefore the SDGs generally are significantly relevant to the developing world (United Nations General Assembly, 2015), we tend to focus our discussion on the small-scale agriculturalists UN agency represent over ninetieth of developing-world farmers (Graeub *et al.*, 2016; International Fund for Agricultural Development/United Nations Environment Programme (IFAD/UNEP), 2013) and conjointly the bulk of these living in economic condition (World Bank, 2015). among that physical and human geographic, one in every of the foremost multifunctional types of agriculture is agroforestry: the combined production of trees and agricultural species on constant piece of land; Garrity *et al.*, 2010; Nair, 1993; Pretty and Bharucha, 2014). Though it's usually been studied for its ecological advantages and peasant-farmer associations (Bhagwat, *et al.* 2008; Horlings and Marsden, 2011; Pretty *et al.*, 2003; Shibu, 2009), scientific proof currently shows that the adoption of agroforestry will increase yields by an element of 2 (average 96% during a multi study review; Pretty and Bharucha, 2014), betting on crop sort, native conditions, and level of experience (Garrity *et al.*, 2010; Pretty and Bharucha, 2014; Pretty *et al.*, 2003; Waldron, Justicia, and Smith, 2015). These yield will increase are shown to replicate multiple system services provided by the trees, as well as increased soil nutrient standing (e.g., through chemical element fixation), reduced crop stress (e.g., through reduced temperature and precipitation extremes), reduced erosion (binding of soil by roots), and regulation of water system (hydraulic uplift of problem by tree roots; Garrity *et al.*, 2010; Leakey, 2014; Pretty *et al.*, 2003; Waldron *et al.*, 2015). What is more, the yield enhancements will be SDG (secure) as a result of agroforestry maintains soil fertility and might even restore degraded lands (Garrity *et al.*, 2010; Leakey, 2014; Nair, 1993). However, food security itself doesn't simply rely upon yield (FAO, 2008; World Bank, 2015).

An extra essential element is resilience to climate change and to shocks (sudden, massive variations in weather, harvests, market costs, and input costs), which might cause major hunger crises (Chappell *et al.*, 2013; Chuku and Okoye, 2009; FAO, 2008; Garrity *et al.*, 2010; McMichael, 2009; Orr and Mwale, 2016; Thorlakson and Neufeldt, 2012; World Bank, 2015). Agroforestry will increase crop resilience to many probably temperature change effects, like drought or higher temperatures, as a result of it enhances water infiltration and storage whereas reducing evaporation and temperature extremes (Charles, Munishi, & Nzunda, 2013; Garrity *et al.*, 2010). It conjointly will increase livelihood resilience as a result of the supply of free ecosystem services by the trees reduces dependence on unpredictable, distant commodity markets; once harvests are poor, the trees conjointly offer different sources of each income and food, as an example, fruit, fodder, or fuel (Charles *et al.*, 2013; Garrity *et al.*, 2010; Thorlakson and Neufeldt, 2012). additionally to advancing food security, agroforestry may also enhance multiple social dimensions of the SDGs.

It provides a pathway out of economic condition (a major driver of hunger itself; World Bank, 2015) as a result of the combination of accumulated yield, low cost, and extra tree-based farm product will considerably increase net farm (Miller, *et al*, 2017; Reyes, *et al*, 2005; Waldron, Justicia, *et a*l, 2012). Additionally, farmer movements within the developing world have expressed a scarcity of equity and dignity within the manner their livelihoods will be negatively laid low with distant supply-chain actors, then have wanted larger local management of food production (e.g., "food sovereignty"; Chappell *et al.*, 2013; Desmarais, 2007; Whitmann, 2011), see also (Cook *et al*, 2015; Thorlakson and Neufeldt, 2012). In agroforests, the reduced dependence on external chemical inputs, and the larger resilience to promote fluctuations, will enhance this sense of management, equity, and dignity in work (Chappell *et al.*, 2013; Thorlakson and Neufeldt, 2012). what is more, on-farm trees generate extensive fuel wood, reducing the necessity to cut down natural forests and conjointly savings mall holder relations (particularly women) from walking long distances (sometimes >20 km) in search of fuel, therefore enhancing women's well-being and liberating them to teach and have a tendency to kids, offer farm labour, or manufacture alternative income (Kiptot, *et al.* 2014; Sharma *et al.*, 2016; Thorlakson and Neufeldt, 2012). The on farm supply of fuel-wood conjointly represents a crucial, hunger-related variety of energy security for rural communities, being low cost, promptly accessible, and basic to extracting decent calories from food (Sharma *et al.*, 2016).

Scenario analysis of causative factors responsible for rural livelihood crisis

There is an urgent need for a realistic scenario analysis of various factors, which contribute to or are linked with the rural livelihoods crisis, for its management by resource restoration-cum-conservation. The scenario analysis of some important factors is recorded here in below.

Land degradation vis-à-vis livelihood scenario

Land resources, the foremost propeller of livelihoods, are subjected to serious qualitative and quantitative degradations, mainly due to faulty land use practices. Rural agrarian livelihoods depend heavily on natural resources, but a vicious circle of over-exploitation and land degradation has set in, which negatively impacts livelihoods. The severe degradation of land, with 57% of the total land area degraded on account of environmental and manmade reasons, such as soil erosion due to water runoff and loss of vegetative cover, over-use of chemicals in agriculture, and more intensive cultivation, is a key concern (IDFC, 2013). The data (The World Bank, 2016) of last fifteen years: 2001 - 2015 reveals just 0.2% increase in agricultural land of the country, which is quite insufficient to support the population boom of the country unless otherwise the waste and fallow lands are put to some productive purpose. Land degradation is an anthropogenic induced process and poses biggest threat to livelihood security of the rural communities across the country. The fast decline in per capita land area, due to inheritances during successive generations, besides unrestrained land acquisitions/diversions for various non-agrarian purposes of the fast growing urbanization, forms a remarkable bottle neck for farming returns. All of these factors combined with increased rate of land degradation contribute towards to livelihood crisis, which can only be resolved by optimization of land productivity through multiple land use.

Agriculture vis-à-vis livelihood scenario

Agriculture, an important base for the rural livelihoods of India, is under serious threats. Since last decade, food insecurity, poverty and climatic adversity have complimented each other hampering livelihoods of the rural poor, destabilizing sustainable development of this country. Indeed, over the last four decades agricultural production has come under increasing risk due to the adverse climate changes, land and soil degradation. Over 70% of the rural households depend on agriculture as their principal means of livelihood (Government of India, 2015). Small holder agriculture faces many challenges including low productivity, high dependence on rain-fed agriculture, insecurity of the traditional land tenure system and environmental degradation due to unsustainable agricultural practices in rural India (Kumar *et al.*, 1999). As such small holder agriculture remains at low productivity leading to high incidence of poverty, low socioeconomic conditions and livelihood stress among rural small holder farmers (Sharma, 1996; Banyal *et al.*, 2011). Large reductions in resilience may mean that the ecosystems, on which livelihoods are based, break down, causing distress (Government of India, 2006). Sustainability of rural livelihoods and resilience of natural resources are two vital coordinates of sustainable development of a country. Escalating demand and supply gaps of agricultural produce promotes deforestation and depletion of forests for sustenance, generating a detrimental poverty-resources degradation vicious circle. As a result of all that, agricultural production has seldom matched the needs of the people adding much to the rural livelihood crisis, which can only be resolved by replacement of conventional farming practices of seasonal and mono-cropping systems with multitier integrated multiple cropping tree-crop plantations under agroforestry, for expeditious optimization of farming output per unit area of land.

Forest vis-à-vis livelihood scenario

Critical livelihood – forest linkage of a huge forest dependent population has been identified as one of the factors affecting forest degradation in India (Forest Survey of India, 2011; Devidar *et al.*, 2010). According to several estimates, India has traditionally been characterized as a low forest cover - low deforestation (LFLD) country exposed to significant direct-human induced deforestation and degradation in past few decades (Devidar *et al.*, 2010 cited in Nayak, *et al.* 2014, Ravindranath *et a*l., 2012). Faster socio- economic progress of forest dwellers (largely living below the poverty line and facing starvation) depends on forestry development (Government of India, 2001). Consequently, India's forests harness a large potential for livelihood based activities for the forest dependent communities, thus bridging the gap between the poor and for established market. With such a huge population depending on forest for subsistence livelihood, the strategies for controlling forest degradation need to be focused on reducing the dependence by creating alternative livelihood opportunities for the forest dependent communities, providing alternative technologies to reduce the gap in demand and supply of forest products by adoption of sustainable harvesting practices. The rural population of India, having majority of marginal and small farmers, constituting around 70% of the total population, are in habit of complementing the agricultural deficits from the forest resources, leading to qualitative and quantitative deteriorations in forest. The forest degradation and diminishing farm productivity, get enflamed by unsustainable dependence of the soaring population, which leads to fury of livelihoods crisis. The recent "Durban Declaration" (WFC, 2015) during the XIV World Forestry Congress, has underlined the importance of forests as fundamental source for food security and improved livelihoods and further added that the forests of the future will increase the resilience of communities by providing food, wood energy, shelter, fodder and fiber; generating income and employment to allow communities and societies to prosper; harboring biodiversity; and supporting sustainable agriculture and human wellbeing by stabilizing soils and climate and regulating water flows. The reports (FAO, 2002,) illustrate that one out of four of the world's poor relies directly or indirectly on forests for their livelihoods. Forest is important natural capital, which combines with other assets to sustain livelihoods, especially among the poor (Warner, 2007). People prefer to generate much of their livelihoods from forests because it is a viable option; however for many, forest dependency is often a last livelihood resort (Bryon and Arnold, 1999). Under such a situation revival of the forest resources, by agroforestry, becomes the need of the era.

Agroforestry vis-à-vis livelihoods linkage scenario

Agroforestry is a viable option for ensuring livelihood security by its multiple usufructs and services. This magnificent land use technology offers an alternative solution to address low land productivity for resource constrained small holder farmers (Puri and Nair, 2004). Agroforestry technologies have extensively been researched and introduced to small holder farmers in India for over three decades (Bijalwan *et al.*, 2011). Agroforestry continues to support livelihood improvement through production of food, fodder, fuel wood and timber and simultaneous mitigation of the adverse impact of climate change (Islam and Quli, 2016). Despite research and extension efforts during last few decades many small holder farmers have still not adopted agroforestry technologies due to many predicaments (Dagar, 2012). In spite of magnificent contribution, the adoption level of agroforestry systems by small farmers is still inadequate (Islam and Quli, 2016). However, unless farmers widely adopt these technologies as part of their farming system, the potential benefits of agroforestry on livelihoods and the environment will not be realized (Ali and Chaturvedi, 2008).

Agroforestry for resolving rural livelihood crisis

At a stage when the ecological censorship has acclaimed global unanimity, the rationale of deliberation on sustainable land-use through agroforestry, becomes very pertinent (Quli and Singh, 2011). A need for diversification in agriculture for improving economy and sustainability has been emphasized (Government of India, 2001). Diversification of agriculture by other land use systems like forestry, animal husbandry, horticulture and fisheries etc., also has been envisaged, to make it more lucrative and ecologically sustainable (Swaminathan, 2001). National Agriculture Policy- 2000 underlines a need for diversification in agriculture with the promotion of integrated and holistic development of rainfed areas on watershed basis and augmentation of biomass production through agro and farm forestry with community involvement. Recessions in national forestry and agricultural sectors have necessitated "multifunctional approach to agricultural farming practices" based on integration of interventions like forestry and animal husbandry into conventional agriculture. Indian Economy has been exposed to challenges due to soaring population, widening demand supply gap of livelihood essentials i.e. food, fodder, fuel, fruit and variety of NTFPs consequent to natural resource degradation and adverse climate change.

Agroforestry having synergistic integration of several production systems has accorded its' proactive adoption around the world, except India, where the pace of adoption are yet to achieve satisfactory level, due to various factors like predominance of small land holdings, inertia of farmers, complications in harvesting and marketing of timber trees, improper marketing facilities etc. The agroforestry has slowly been gaining ground in India and as per data (Forest Survey of India, 2013,) an area of 111,554 Km^2 (3.39%of the total geographical area of the country) of the country, is now under agroforests, which have initiated bio-remediation of the socio-economic and eco-climatic crisis, for reverting rural livelihood sustainability. According to the 2019 report, the total forest cover of the country is 712,249 square kilo meters (21.67 percent of India's total geographical area) slightly up from 708,273 sq. km (21.54 percent) in 2017. The tree cover of the country is 95,027 sq. km (2.89 percent of the total area) again slightly up from 93,815 sq. km. (2.85 percent) in 2017. With a proven credential to boost livelihoods by augmenting agricultural risks, agroforestry enriches the soil by fixing nitrogen by leguminous trees, improves drainage, expedites soil enrichment by steady and efficient nutrient cycling, optimizes land productivity to enhance farming returns. Minimization of economic risks by high income generation and in cropping enterprises is one of the most prominent benefits of agroforestry. It provides long term investment opportunity, diversified land use, commercial tree cropping and best option for absentee landlords besides generating diversified on-farm employment, wood and Non-Wood Forest Produce (NWFP) and raw-material supply to forest based industries. Agroforestry supports large number of forest based industries besides offering lucrative direct returns and employment generation.

Agroforestry systems for revival of sustainable livelihoods

Properly designed agroforestry systems generate enormous potential to revolutionize the productivity of bio-production systems, on sustainable basis, in terms of qualitative as well as quantitative enhancements in production of food grains, fruits, fodder, industrial raw materials, fish, egg, chickens, red meat, milk, silk, lac and various other animal products, as well as variety of NWFPs for resurgence of sustainable livelihoods with very effective and far reaching indispensable ecosystem ameliorative impact. Based upon the global analysis of agriculture and water status, adoption of multi sectoral integrated land-use with long lasting solutions has been recommended (Bossio *et al.*, 2007) obviously suggesting use of agroforestry under Farming systems. This paper mainly targets rehabilitation of the total 67.81 million ha, of wastelands, which amounts to 20.62% of the total geographical area of the country, across the country. However, the temporary/ occasional fallows and conventional farm land with low/ inadequate farming intensity are also recommended to be covered by agroforestry under this proposed plan. The important technical details like prominent agroforestry systems, tree-crop combinations, preferred tree under the various agroforestry systems being declared suitable and multipurpose tree species (MPTs) priority in various zones are reproduced from the authentic records (Anonymous, 2001), to facilitate appropriate decision making by the stakeholders of agroforestry.

Women's Participation in Agroforestry

Some females were found to be involved in agroforestry as sole farmer (19.1%), thus agroforestry had improved their social and economic status. It was found that 33% of female agroforestry farmer had been widowed, so they become involved in agroforestry to feed themselves and their families. About 67% of married females had elderly and frail husbands (none had never married) took over the farming tasks. Both male and female family members play a critical role in agroforestry by working together in tree improvement and propagation. Women share the responsibility with men for tree and crop protection, irrigation, weeding and harvesting of annual crops. Women are also responsible for the post harvest storage and processing of food products, e.g. for homemade food from fruits, vegetables and cereal crops for both household consumption and for the market.

Livelihood Security by Agroforestry

Food Security

Asia is the "continent of the current century", according to many; yet, some analysts have shown that many Asian countries may not be able to feed their projected populations in the 21st century. On the one hand, there is less land per person in Asia today than in other parts of the world6 and on the other, productive land is progressively being displaced by urbanization. Historically, food production in the overall Asian context increased at the same rate as that of human population FAO. However, population growth has out maneuvered the food production trends in the past decade, implying the need to augment food production. According to FAO, there are about 800 million people in the developing world who suffer from hunger. And most of this (60%) is in Asia with South Asia accounting for about 36%. To make matters worse, increases in cereal yields are slowing down in all regions of the world due to the so called "technology fatigue", and Asia is no exception. Woody perennial based production systems, such as agroforestry, have the potential to meet the food security of people (Sahoo and Wani, 2019). Farmers depend more on annual crops, the small and marginal farmers in the tropics have long been practicing agroforestry to meet their food, fodder and fuel requirements. Apart from ensuring food production, such systems also would enhance economic returns to the growers. Consistent with this, Rasul and Thapa in a case study of the degraded agricultural lands of Chittagong Hill Tracts (Bangladesh) reported that economic returns from agroforestry were greater than that from jhum. The higher cash incomes provide greater "buying power" with respect to food, especially when agriculture is not practiced, or when the crops fail. Moreover, diversified production is a form of risk avoidance, which is of special relevance in the context of the current agricultural crises that many countries in South and Southeast Asia are experiencing. Agroforestry to provide alternate sources of income and employment to the rural poor also has been highlighted. The diverse products (fruits, vegetables, spices etc.), which are available year-round in systems such as home gardens not only contribute to food security during the "lean" seasons but also ensure food diversity. They are also sources of mineral nutrients for improving household nutritional security especially for "at-risk populations" (e.g., women and children). In experimental studies, target families significantly increased year-round production and consumption of vitamin-rich fruits and vegetables compared to a control group without gardens. This, in turn, alleviated deficiencies of iodine, vitamin A, and iron and made children of garden owners less prone to xerophthalmia. As little or no chemical inputs are used, the produce from agroforestry is also expected to be of superior quality. Over the period when input usage in agriculture was promoted in Asian agriculture, agroforestry being less input intensive was overlooked as a means of food production. The development community, in particular, was not fascinated by such mixed gardens with scattered and/or boundary planted trees. The woody perennial based mixtures were also thought to be less productive and difficult to manage; instead, the "replicable models" of input intensive production practices became fashionable. The smallholder mixed tree-gardens in Asia thus represent a substantial unexploited potential for enhancing productivity and profitability. Beside that households food security condition in Jassore district of Bangladesh highly improved by following agroforestry.

Poverty reduction

Agroforestry provide a greater contribution of the total income of farmers per year. This contribution is obtained from agricultural crops, forestry (timber) and livestock. Agricultural crops such as cocoa, coffee, cloves, rice and fruits derive most of their income due to crop harvest to include plants that do not require a long time and has economic value so that farmers get a continuous income to meet daily needed. Timber species are widely grown in agroforestry is chrysolite, bayur, teak, sengon, medang and hibiscus. Timber grown mostly for long-term savings, if households need large amounts of cash then the wood is cut down. Timber prices vary widely depending on the type, age, size, and quality of the wood. Commercialize livestock farmers as savings for the future. Many households keep cattle that are regularly sold or redeemed for cash and food as part of their normal activities yearly. Cows and goats are the animals that most commonly cultivated by farmers in addition to the use of manure as a fertilizer for crops as well as the fuel of biogas. Cost of production in agroforestry management covers the cost of fertilizer, pesticide, labour, and seed. Land management is not carried out intensively. By following agroforestry as suggested above farmers get additional income that would help to expenditure and ultimately involved to poverty reduction. People having more physical assets reveal that he/she enjoys more social status than others. They observed during their study farmers which are performing agroforestry having more no. of physical assets as compare to non agroforestry practitioner (Table1).

Table 1: Physical Asset of the Respondents

Physical asset	**Agroforestry Practitioners (Percentage)**	**Non agroforestry**
Practitioners (Percentage)		
Television (No.)	55	38
Radio (No.)	8	14
Mobile Phone (No.)	100	95
Bicycle (No.)	74	66
Motorcycle (No.)	24	10
Power-tiller (No.)	12	5
Spray-machine (No.)	26	32
Shallow-machine (No.)	21	15
Paddy threshing machine (No.)	33	20

The concept of Trees outside Forests (ToF) emerged in the early 90s FAO as a holistic approach which encompasses integrated tree based farming system in farm lands and pasture lands to promote sustainable agricultural production and forest resource conservation. Indeed, in small-scale agricultural production systems, TOF management seems to hold a high promise as a bridge between food production and environmental protection, due to its capacity to restore the ecosystems and improve soil fertility. Farmers welcome tree cropping because they are economically advantageous since they provide substantial cash incomes which could be recycled into food in case of crop failure. Thus agroforestry as a strategy to uplift the economic conditions of the farmers while rehabilitating the degraded uplands has made inroads in the productivity site. Although the future returns from the harvesting of mature trees in the farm forests are expected to further improve the farmers' income and well being.

Environmental Security

Enhancing soil fertility: The primary objective of soil conservation is to improve or maintain soil fertility. To achieve this, control of erosion, maintenance of organic matter and physical properties, organic matter addition, maintenance of nutrient is necessary. In this way agroforestry system constitute sustainable land use and helps to improve soils in the number of ways. Maintenance and enhancement of soil fertility vital for global food security and environmental sustainability. Ecologically sound agroforestry systems such as intercropping and mixed arable-livestock systems can increase the sustainability of agricultural production while reducing on-site and off-site consequences and lead to sustainable agriculture. Alternate land-use systems such as agroforestry, agro-horticultural, agro-pastoral and agro silvi-pasture are more effective for soil organic matter restoration. It was also observed that soil organic carbon status increased by 5 to 6 times higher in agroforestry system than growing of either sole tree or sole crop.

Biodiversity conservation: Over exploitation of natural resources is a major challenge for sustainable production and livelihood security. Deforestation is that major cause which affected the biodiversity of an ecosystem. Agroforestry with components like trees, agricultural crops, grasses, livestock etc. provides all kinds of life support. However, agroforestry may not entirely reduce the deforestation but in many cases it acts as an effective buffer to deforestation. Trees in agroforestry system act as a refuse to biodiversity after catastrophic events such as fire. The traditional societies of coastal belts and tropics of the country are practicing home gardens and sacred groves which helped in biodiversity conservation.

Carbon sequestration: Tree components in agroforestry systems can be significant sink of atmospheric carbon (C) due to their fast growth and high productivity. By including trees in agricultural production systems, agroforestry can, arguably, increase the amount of C stored in lands devoted to agriculture, while still allowing for the growing of food crops26. In agroforestry system, tree components are managed, often intensively by pruning for minimizing competition and maximize complimentarily. The pruned materials are mostly non- timer products. Such materials are often returned to soil. Besides, the amount of biomass and therefore C that is harvested and exported from the system is relatively low in relation to the productivity of the tree. Therefore, unlike in tree plantations and other mono culture systems, agroforestry seems to have unique advantage in terms of C sequestration. In India, evidence is now emerging that agroforestry systems are promising land use system to increase and conserve aboveground and soil C stocks to mitigate climate changes. The average potential of agroforestry has been estimated to be 25 t carbon ha^{-1} over 96 m ha. In this way the total potential of agroforestry in India to store C is about 2400 t, but the C storage capacity varied from region to region and also depends upon the growth and nature of tree species involved in the system.

Livelihood Security Comparison of Traditional Agroforestry System and Commercial Agroforestry System

In the traditional agroforestry systems since the trees are naturally growing especially in traditional agroforestry region and are just allowed to be thriving by the farmers, the costs associated with management of the trees are negligible except that of indirect costs associated with the shade and competition due to moisture and nutrient needs. Therefore, only the benefits from trees on account of harvest and sale of tree produce were accounted, while commercial agroforestry system is characterized by trees in close association with crops either on farm bunds/ boundaries or within the fields. Socio-economic diagnosis of traditional as well as commercial agroforestry practices followed by farmers in western Uttar Pradesh carried out by Dwivedi et al., and they found that tree species like *Azadirachta indica, Acacia nilotica, Dalbergia sissoo* and *Eucalyptus* spp. were dominant species in traditional system whereas, *Populus deltoides* and Eucalyptus spp. were the main species of commercial agroforestry. Fuel wood (50.6 %) was major driving force for agroforestry adoption followed by additional income (24.4 %) and shade (17.5 %) in traditional agroforestry region indicated in table While, additional income (71.3 %) was the major factor in commercial agroforestry region. Although traditional agroforestry seems less promising as compared to commercial agroforestry, but it is also relevant to the farmers (Table-2).

Table 2: Determinants of traditional Vs commercial agroforestry system

Traditional agroforestry system		**Commercial agroforestry system**	
Major reason	**Percentage**	**Major reason**	**Percentage**
Additional income	71.3	Fuel wood	50.6
Source of money in emergency	17.5	Additional income	24.4
Source of fuel wood	2.5	Shade	17.5
Source of employment	4.4	Timber	3.8
Others	4.4	Others	3.8

Recommendations for Promoting Agroforestry

Keeping in view the magnificent potentials of agroforestry in restoration of eco climatic and socio-economic regimes of rural areas, which uphold the rural livelihoods, following recommendations (Government of India, 2001) are to be taken care of:

- Commercial agroforestry should be practiced in areas where irrigation is available. Suitable species include *Acacia nilotica*, Bamboo species, *Casuarina equisetifolia*, *Eucalyptus* species, *Populus deltoides* and *Prosopis cineraria* for different agro- climatic and edaphic conditions;
- High-tech modern nurseries should be established on catchment area basis
- Agroforestry should be tried in rain fed areas by adopting various agroforestry models that would provide additional income as well as complementing agriculture
- Elite clones (higher yielding and disease resistant) of important agroforestry species should be developed for different edaphic and climatic conditions. Corporate/ Private sector should also be encouraged to take up R & D activities and promote new agroforestry products

- Agroforestry product research, new product development, new designs and quality standards should be evolved for downstream processing
- Market information system should be developed to inform farmers about the major buyers, market trends, etc.
- All restrictions on felling of trees, logging, transport and marketing of forest produce should be removed
- Establishment of agroforestry boards and marketing federations should be encouraged to improve the bargaining power of farmers in domestic and export market.

Financing forests' contribution to climate change mitigation and adaptation sources of finance have emerged to support forest-related mitigation efforts, together with the clean development mechanism (CDM) below the city protocol and voluntary carbon markets, and, recently, the REDD + (Reducing Emission from Deforestation and forest Degradation in Developing Countries) Partnership, the Forest Carbon Partnership Facility (FCPF) and also the Forest Investment Program (FIP) of the Strategic Climate Fund. REDD+ activities also are actively supported by bilateral and organization funding. Numerous funds managed by the global setting Facility (e.g. the variation Fund), also as different support – through tripartite, bilateral and NGO channels, offer money help for global climate change adaptation of forests, biological science and forest- dependent individuals. The necessity for funding for adaptation measures is growing, as recognized within the Copenhagen Accord, which places equal weight on mitigation and adaptation.

Overcoming Challenges

Beneficial effects are not universal and certain practices may predominate negative or neutral effects and the performance of agroforestry systems depends on relative influence of tree species selection and management, soil characteristics, topography, rainfall, agricultural practices, priority for food security, economic and social development options, etc. (Kumar, 2006; Mbow *et al.*, 2014 b). Firstly, to achieve desired social and environmental benefits designing the right agroforestry system by selecting the most suitable plants and management practices is a must. (Mbow *et al.,* 2014 a). Selection of species must carefully done in response to local priorities and biophysical conditions (Mbow *et al.*, 2014 a). Even though, the shade trees are competitors for water use, some studies have shown that by understanding the different root attributes of intercropped trees such as contrasting spatial rooting pattern, root morphology and mycorrhizal status it is possible to achieve optimal use of complementary resources such as water (Tscharntke *et al.*, 2011). For an example, use of a combination of deep rooted and shallow rooted species in agroforestry systems can improve the efficient use of soil water and nutrients. Not only the appropriate crops and trees, favourable sites, suitable management practices, it is also important to integrate those practices into local livelihood systems and communities as they act as initiators of the transformative change inland use. Asagroforestry takes relatively longer period to realize benefits than other conventional agricultural systems, farmer involvement should be stimulated by the assuring land security and tree tenure. Land should be recognized as a common benefit while applying community inclusive stewardship principles. Continuous low returns may negatively affect the capability of farmers to improve their optimal management systems (Jamnadaas *et al.*, 2013; Mbow *et al.*, 2014 a; Ofori *et al.*, 2014; Place *et al.*, 2012). Hence, improving micro-credit at local level, appropriate and adequate financing at grass-root level and promotion of agroforestry as a potential and profitable investment is also crucial.

To meet the desired address social and economic barriers, uphold social conditions such as gender synergies, development of more coordinated, structured and fair market systems with collective bargaining for both inputs and outputs of agroforestry, better transport infrastructure, involvement of lesser intermediaries in the supply chain, better investment in characterization of tree foods, provision of incentives from payment for eco system services and certification schemes should be adopted (Garitty, 2004; Leakey, 2010; Jamnadaas *et al.*, 2013; Tscharntke *et al.*, 2011). In addition, management of the demand side aspects such as change in diet specially in the growing urban population percentage should also considered seriously (Mbow *et al.*, 2014 b). Another major challenge is to scale up successful tree domestication approaches (Ofori *et al.*, 2014). Improving the knowledge beneficial traits of the existing indigenous plants used in agroforestry and development of new cultivars from those indigenous species, availability and ample access to high quality planting material and the safe trans-boundary transport of superior cultivars developed in different parts of the world are also important. Furthermore, the small holder farmers play an undeniable role in the agriculture sector. In South Asia about 80% agricultural lands are less than 0.6 ha in extent (Kumar, 2006).

Failure to extend advance agricultural methods hinders the small holder farmers in developing countries in achieving optimal benefits in agroforestry. Apart from the innovative limitations, small holder farmers suffer structural limitations due to the limited investment opportunities available compared to other conventional agricultural practices such as monocultures of cash crops. In general, profitability can be increased through improving and diversification of output per unit area of tree/crop/livestock and by addition of new products to enhance the financial diversity and flexibility. Hence, attention of scientific community should relate to methods and tools appropriate to gain maximum benefits from agroforestry in various human and ecological settings. Also, scientific knowledge should be translated to decision makers and the policy developers and implementers should abandon the conservative segregation of environment, agriculture and other related sectors as separate domains and embrace the need of a holistic approach to manage all the aspects under one system. Since most ecological and economic studies are conducted separately the information is difficult to be linked and it is challenging to estimate real time financial benefits, extent and trends of agroforestry, as government inventories do not include those (Morgan *et al.*, 2010). Even Food and Agricultural Organisation's Global Forest Assessment do not recognize agroforestry as a land use as it is categorized as agriculture or as a forest. However, available information hints the percentage of agroforestry is increasing many parts of the world (Mbow *et al.*, 2014 a). The growing appreciation in developed countries will enhance understanding and support for its expansion in the developing countries ensuring the needed investments (Garrity, 2004).

Conclusion

Agroforestry have potential to contribute to the maintenance of biodiversity in natural systems due to the reduction in overreliance of rural communities on natural forest resources, as they are able to maintain their production systems through improved agroforestry systems. Commercial agroforestry important for assured income as compared to traditional, but both forms of agroforestry have specific roles to play in the livelihoods. We now have the opportunity to refocus both policy and finance, so that they better reflect an appropriate balance of agricultural solutions in a multi goal framework. For less-mainstream techniques like agroforestry, such a rebalancing would bring large new investment, research, and institutional improvements, driving yields upwards. The needs of the poor must be directly addressed with their participation to ensure sustainability on which steady growth depends. Improving livelihoods will need that new approaches are adopted and financial commitment secured. By assessing and capitalizing on this potential, we could significantly enhance global goals on food security, social well-being, and environmental integrity, as we have committed to under the SDGs. Agroforestry allows the growth of multiple crops simultaneously and provides several livelihood benefits to farming households. Commercial opportunities exist for farm communities to transform their traditional agroforestry systems towards market orientation. To achieve this transformation, smallholder farmers must develop intensive deliberate management systems designed to yield quality products from priority species that meet market specifications. Most farmers are ill-prepared for this challenge because their traditional extensive management approach produces small quantities of many products primarily for household consumption with limited market sales.

References

Ali, M.S., & Chaturvedi, O.P. (2008). Pattern and constraints of adoption of agroforestry systems in north Bihar. Journal of Tropical Forestry, 24(1/2),10-15.

Amundson, R., Berhe, A., Hopmans, J. W., Olson, C., Sztein, A., & Sparks, D. L. (2015). Soil and human security in the 21st century. Science, 348,6235.

Banyal, R., Masoodi, N. A., Masoodi, T. H., Sharma, L. K., & Gangoo, S.A. (2011). Knowledge and attitude of farmers towards agroforestry practices in North Kashmir- a case study. The Indian Forester, 137(12), 1377-1381.

Bélair C., Ichikawa K., Wong B.Y.L., and Mulongoy K.J. (eds.), (2010) : Sustainable use of biological diversity in socio-ecological production landscapes, Background to the 'Satoyama Initiative for the benefit of biodiversity and human well-being', Secretariat of the Convention on Biological Diversity, Montreal, Technical Series no. 52, pp.184.

Bhagwat, S., Willis, K., Birks, H., and Whittaker, R. (2008). Agroforestry: Arefuge for tropical diversity? Trends in Ecology and Evolution, 23,261–267.

Bijalwan, A., Sharma, C. M., & Kediyal, V. K. (2011). Socioeconomic status and livelihood support through traditional agroforestry systems in hill and mountain agro ecosystem of Garhwal Himalaya, India. The Indian Forester, 138(12), 1423-1430.

Bossio, D., Critchely, W., Geheb, K., Lynden, G. V, & Mati, B. (2007). Comprehensive assessment of water management in agriculture. In D. Molden (Ed.), Water for food, Water for life. London, UK: Earthscan and International Water Management Colombo: Institute.

Byron, N., & Arnold, M. (1999). "What futures for the people ofthe tropical forests?",World Development, 27(5), 789 –805.

Chappell, M. J., & LaValle, L. A. (2009). Food security and biodiversity: Can wehave both? An agroecological analysis. Agriculture and Human Values, 28, 3–26.

Chappell, M. J., Wittman, H., Bacon, C. M., Ferguson, B. G., Barrios, L. G., Barrios, R. G., & Perfecto, I. (2013). Food sovereignty: An alternative paradigm for poverty reduction and biodiversity conservation in Latin America. F1000 Research, pp. 61.

Charles, R., Munishi, P., & Nzunda, E. (2013). Agroforestry as adaptation strategy under climate change in Mwanga District, Kilimanjaro, Tanzania. International Journal of Environmental Protection, 3,29–38.

Chuku, C. A., &Okoye, C. (2009). Increasing resilience and reducing vulnerability in sub-Saharan African agriculture: Strategies for risk coping and management. African Journal of Agricultural Research Special Review, 4,1524–1535.

Cook, S., Silici, L., Adolph, B., & Walker, S. (2015):Sustainable Intensification revisited. London, England: IIED.

Dagar, J. C. (2012). Utilization of degraded lands/ habitats and poor quality water for Livelihood security and mitigating climate change. Indian Journal of Agroforestry, 14(1),1-16.

Davidar, P., Sahoo, S., Mammen, P. C., Acharya, P., Puyravaud, J. P., Arjunan, M., Garrigues, J. P., & Roessingh, K. (2010). Assessing the Extent and Causes of Forest Degradation in India: Where do we Stand? Biological Conservation, 43(12), 2937– 2944.

Desmarais, A. (2007): La via campesina: Globalization and the power of peasants. London, England: Pluto Press.

FAO.(2002) : Forest-based poverty reduction: A brief review of facts, figures, challenges and possible ways forward. Paper prepared for the International Workshop on "Forests in October, 2002, Tuusula, Finland.

FAO (2008). Food and Agriculture Organization of the United Nations (2008). An introduction to the basic concepts of food security. Rome, Italy.

FAO (2015) : Food and Agriculture Organization of the United Nations, International Fund for Agricultural Development, & World Food Programme (2015). The state of food insecurity in the world 2015. Meeting the 2015 international hunger targets: Taking stock of uneven progress. Rome, Italy: FAO.

Foley, J. A., Ramankutty, N., Brauman, K. A., Cassidy, E. S., Gerber, J. S., Johnston, M., & Zaks, D. P. M. (2011): Solutions for a cultivated planet. Nature, 478

Forest Survey India (FSI) (2003) : State of Forest Report 2003. Forest Survey of India, Ministry of Environment and Forests, Dehradun.

Forest Survey of India. (2011) : India State of Forest Report. Ministry of Environment & Forests, Government of India, Dehradun. Retrieved from http://fsi.nic.in/cover_2011/ trees in agroforestry. pdf.

Forest Survey of India. (2013). India State of Forest Report. Ministry of Environment & Forests, Government of India,Dehradun.

Garrity D.P., (2004) : Agroforestry and the achievement of the MillenniumDevelopment Goals, Agroforestry System,s 61,5–17.

Garrity, D. P., Akinnifesi, F. K., Ajayi, O. C., Weldesemayat, S. G., Mowo, J. G., Kalinganire, A., Bayala, J. (2010): Evergreen agriculture: A robust approach to sustainable food security in Africa. Food Security, 2,197–214.

Godfray, H., Beddington, J., Crute, I., Haddad, L., Lawrence, D., Muir, J. F., Toulmin, C. (2010): Food security: The challenge of feeding 9 billion people. Science, 327,812–818.

Godfray, H. C. J., & Garnett, T. (2014): Food security and sustainable intensification. Philosophical Transactions of the Royal Society of London. Series B, Biological Sciences, 369,20120273.

Graeub, B. E., Chappell, M. J., Wittman, H., Ledermann, S., Kerr, R. B., & Gemmill Herren, B. (2016): The state of family farms in the world. World Development, 87, 1– 15.

Griggs, D., Stafford-Smith, M., Gaffney, O., Rockstrom, J., Ohman, M. C., Shyamsundar, P., Noble, I. (2013): Sustainable development goals for people and planet. Nature, 495, 305–307.

Government of India. (2001) : Report of the Task Force on Greening India for Livelihood Security and Sustainable Development. Planning Commission Government ofIndia.

Government of India. (2006) : National Environment Policy, Ministry of Environment, Government of India.

Government of India. (2015) : Pocket Book on Agricultural Statistics. Ministry of Agriculture & Farmers Welfare Department of Agriculture, Cooperation & Farmers Welfare Directorate of Economics & Statistics, New Delhi.

Horlings, L. G.,& Marsden, T. K. (2011): Towards the real green revolution? Exploring the conceptual dimensions of a new ecological modernisation of agriculture that could ''feed the world.''. Global Environmental Change, 21, 441–452.

International Assessment of Agricultural Science & Technology for Development(2013). Agriculture at a cross roads. Washington, DC: Island Press.

International Fund for Agricultural Development (IFAD)/United Nations Environment Programme (UNEP). (2013). Smallholders, food security and the environment.

IDFC. (2013) : India Rural Development Report 2012|13. Rural Development Network, Infrastructure Development Finance Company Limited, Foundation, ORIENT Blackswan Private LIMITED, pp.280.

Islam, M. A., & Quli, S. M. S. (2016) : Motivation strategy for agroforestry intensification among small holders. Advances in Life Sciences, 5(10),3878-

Kiptot, E., Franzel, S., & Degrande, A. (2014): Gender, agroforestry and food security in Africa. Current Opinion in Environmental Sustainability, 6,104–109.

Kumar, A. A., Babu, B., & Ramachandran, V. (1999). Attitude of farmers towards Agroforestry programmes in Kerala. Indian Journal of Forestry, 22(2), 155-159.

Leakey R. R. B., (2010) : Should We be Growing More Trees on Farms to Enhance the Sustainability of Agriculture and Increase Resilience to Climate Change?, A Special Report, International Society of Tropical Foresters, USA, pp.1-12.

Leakey, R. R. (2014): The role of trees in agroecology and sustainable agriculture in the tropics. Annual Review of Phytopathology, 52,113–133.

Maxwell, S. L., Fuller, R. A., Brooks, T. M., & Watson, J. E. (2016): Biodiversity: The ravages of guns, nets and bulldozers. Nature, 536,143–145.

Mbow C., Noordwijk M., Prabhu R. and Simons T., (2014 a) : Knowledge gaps and research needs concerning agroforestry's contribution to Sustainable Development Goals in Africa, Current Opinion in Environmental Sustainability, 6, 162–170.

Mbow C., Smith P., Skole D., Duguma L. and Bustamante M., (2014 b) : Achieving mitigation and adaptation to climate change through sustainable agroforestry practices in Africa, Current Opinion in Environmental Sustainability, 6, 8–14.

McMichael, P. (2009): A food regime analysis of the ''world food crisis.''.Agriculture and Human Values, 26,281–295.

Miller, D. C., Munoz-Mora, J., & Christiaensen, L. (2017): Prevalence, economic contribution and determinants of trees on farms across Sub-Saharan Africa. Forest Policy and Economics.

Montgomery, D. R. (2007): Soil erosion and agricultural sustainability. Proceedings of the National Academy of Sciences of the United States of America, 104, 13268–13272.

Morton, J. (2007): The impact of climate change on smallholder and subsistence agriculture. Proceedings of the National Academy of Sciences of the United States of America, 104,19680–19685.

Morgan, J. A., Follett R .F., Allen L. H., Del Grosso S., Derner J. D.,Dijkstra F., Franzluebbers A., Fry R., Paustian K. and Schoeneberger M. M., (2010) : Carbon sequestration in agricultural lands of the United States, Journal of Soil and Water Conservation, 65(1), pp. 6A-13A.

Nair P. K.R., (1993) : An Introduction to Agroforestry, International Centre for Research in Agroforestry, Kluwer Academic Publishers, The Netherlands, pp.499.

Nayak, B. P., Kohli, P., and Sharma, J. V. (2014) : Livelihood of local communities and forest degradation in India: Issues for REDD+. A policy brief.

Ofori D. A., Gyau A., Dawson I. K., Asaah E., Tchoundjeu Z. and Jamnadass R., (2014) : Developing more productive African agroforestry systems and improving food and nutritional security through tree Domestication, Current Opinion in Environmental Sustainability, 6,.123–127.

Orr, A., & Mwale, B. (2016): Adapting to adjustment: Smallholder livelihood strategiesin southern Malawi. World Development, 29,1325–1343.

Potts, S. G., Biesmeijer, J. C., Kremen, C., Neumann, P., Schweiger, O., & Kunin, W. E. (2010): Global pollinator declines: Trends, impacts and drivers. Trends in Ecology and Evolution, 25,345–353.

Pretty, J. (2008): Agricultural sustainability: Concepts, principles and evidence. Philosophical Transactions of the Royal Society of London. Series B, Biological Sciences, 363,447–465.

Pretty, J., & Bharucha, Z. P. (2014): Sustainable Intensification in Agricultural systems. Annals of Botany, 114, 1571–1596.

Pretty, J. N., Morison, J. I., & Hine, R. E. (2003): Reducing food poverty by increasing agricultural sustainability in developing countries. Agriculture, Ecosystems & Environment, 95,217–234.

Puri, S., &Nair, P. K. R. (2004) : Agroforestry research for development in India: 25 years of experiences of a national program. Agroforestry Systems, 61/62, 437–452.

Quli, S. M. S., & Singh, C. S. (2011) : Agroforestry under integrated farming systems. In A. K. Sarkar, R. S. Singh, M. S. Yadav & C. S. Singh (Eds.), Integrated farming systems: For sustainable production (pp. 294-329). Udaipur: Agrotech Publishing Academy.

Ravindranath, N.H., Srivastava, N., Murthy, I.K., Malaviya, S., Munsi, M., & Sharma, N. (2012) : Deforestation and forest degradation in India implications of REDD+. Current Science, 102(8), 1-9.

Reyes, T., Quiroz, R., & Msikula, S. (2005): Socio-economic comparison between traditional and improved cultivation methods in agroforestry systems, East Usambara Mountains, Tanzania. Environmental Management, 36, 682–690.

Sahoo, G.R. and Wani, A.M. (2019): Multifunctional Agroforestry Systems in India for Livelihoods. Annals of Horticulture, 12 (2) :139-149.

Sharma, K. K. (1996) : Agroforestry in farming systems development. The Indian Forester, 122(7),547-599.

Sharma, N., Bohra, B., Pragya, N., Ciannella, R., Dobie, P., & Lehmann, S. (2016): Bioenergy from agroforestry can lead to improved food security, climate change, soil quality, and rural development. Food and Energy Security, 5,165–183.

Stern, N. (2007): The economics of climate change: The Stern review. Cambridge, England: Cambridge University Press.

Swaminathan, M. S. (2001, Jan 11) : Shaping our agriculture future. The Hindu, Online edition of India's National News Paper.

The Government Office for Science (2011): Foresight. The future of food and farming. Final project report. London, England.

The Royal Society (2009): Reaping the benefits. Science and the Sustainable Intensification of global agriculture. London, England.

The World Bank. (2016): Agricultural land (% of land). Retrieved from http://data.worldbank.org/indicator/ AG.LND.AGRI.ZS.

Tan, Z. X., Lal, R., & Wiebe, K. D. (2005): Global soil nutrient depletion and yield reduction. Journal of Sustainable Agriculture, 26,123–146.

Tscharntke T., Clough Y., Bhagwat S. A., Buchori D., Faust H., Hertel D., Holscher D., Juhrbandt J., Kessler M., Perfecto I., Scherber S., Schroth G.,Veldkamp E. and Wanger T. C., (2011): Multifunctional shade-tree management in tropical agroforestry landscapes–a review, Journal of Applied Ecology, 48, 619–629.

UNDP, India. (2012) : Greening Rural Development in India, Vol-I", United Nations Development Program, New Delhi, pp.61.

Thorlakson, T., & Neufeldt, H. (2012): Reducing subsistence farmers' vulnerability to climate change: Evaluating the potential contributions of Agroforestry in western Kenya. Agriculture & Food Security, 1,15.

Tscharntke, T., Clough, Y., Wanger, T. C., Jackson, L., Motzke, I., Perfecto, I., Whitbread, A. (2012): Global food security, biodiversity conservation and the future of agricultural intensification. Biological Conservation, 151,53–59.

United Nations General Assembly. (2015): Transforming our world: The 2030 Agenda for sustainable development, A/RES/70/1.

Waldron, A., Justicia, R., & Smith, L. (2015): Making biodiversity friendly cocoa pay: Combining yield, certification, and REDD for shade management. Ecological Applications, 25, 361–372.

Waldron, A., Justicia, R., Smith, L. E., & Sanchez, M. (2012): Conservation through Chocolate: A win-win for biodiversity and farmers in Ecuador's lowland tropics. Conservation Letters, 5,213–221.

Whitmann, H. (2011). Food sovreignty: A new rights framework for food and nature? Environment and Society, 2, 87–105.

World Bank (2015): Ending poverty and hunger by 2030. An agenda for the global food system. Washington, DC:Author.

World Forestry Congress. (2015): Press release: World Forestry Congress sets out vision for future of forests, 11 September, Durban, South Africa.

26

Nutritional Security and Livelihood Upliftment of Tribals Through Tuber Crops

N. Jena and S.K. Mohanty

Introduction

Agriculture is the measure livelihood of tribals in India. Indigenous Tuber crops, like Elephant foot yam (*Amorphophallus*) and *Dioscorea* play a valuable role to uplift the socio and economic status of people of tribal areas (tribal farmers) in terms of nutritional security and livelihood. In every year rural India subjected to wide spread of poverty, malnutrition, infant mortality rate etc. Ecological degradation, climate change, erratic rainfall and drought are the principal causes of food insecurity in these tribal areas resulting migration, periodic starvation and deaths. Therefore under erratic climate, tuber crops like *Amorphophallus* and *Dioscorea*, played a crucial role as life support crops to combat hidden hunger. Tuber crops are the traditional crops cultivated in order to meet the food requirements and give economical sustainability to the people. After cereals and legumes, tuber crops constitute a major staple or subsidiary food for all. *Amorphophallus paeoniifolious* and *Dioscorea alata*, important tropical tuber crops are excellent dietary staple for low-income consumers in tribal areas of Odisha state, better opting tuber crops cultivation rather than forest produces. These tubers are known to supply cheap source of energy, especially for weaker sections of the population in the context of food and nutritional security, and possess good capacity to withstand adverse abiotic stresses (Edition. S., 2006).Tuber crops play a vital role as dietary energy source in the human diet, having various nutritional and health benefits such as anti-oxidative, hypoglycemic, antimicrobial and immunomodulatory activities (Chandrasekara *et al*., 2016).

Elephant foot yam (EFY) tubers commonly consumed as vegetables, are good source of carbohydrates, proteins, fibers and minerals but low in fat. Secondary metabolites of this tuber have a significant antiglycemic effect and safely recommended to diabetic patients (Rahman, 2020). The tuber also contains anti-nutrient factors such as oxalate and phytate which create itching while eating. However, a high yielding local selection nonacrid variety 'Gajendra' of Andhra Pradesh was released for all India cultivation. The nutritional composition of *Dioscorea alata* consists of carbohydrates (starch & sugar), proteins, ash and fibres could be good dietary source to its consumers (Baah et. al, 2009), is also documented for their therapeutic properties. The tubers of these crops used in many indigenous medicinal preparation for various ailments (Mishra and Sriram,2002). *Amorphophallus paeoniifolious* is raised and required 8-9 months to harvest, when till yellowing of leaves and complete drying up of the whole plants, indicated its harvesting time. *Dioscorea alata* leaves turned total yellow and plants starts drying up at maturity, required 7-8 months to harvest. Both long duration crops are well grown in the fertile sandy loam soil of pH 5.5-7.0 and pH 6.0-6.5, with yield of 30-40t/ha and 25-35t/ha respectively. But it can be grown in all soil types at pH range of 5.5 to 8.0 with comparatively low yield. These crops can withstand long periods of drought and low fertility levels. Physiologically matured tubers are generally harvested depending upon the market demand and own food need, as their dormancy period varies from 45-65days.

Amorphophallus and *Dioscorea* both are suitable for mixed farming with crops like millets, maize, arhar, turmeric, red gram and some vegetables, also successively inter-culture with coconut, mango, pomegranate, guava, banana, papaya, sapota etc. with low investments and maintenance. Intercropping has several advantages as better utilization of environmental factors, greater yield of food, increasing the return per unit area and insurance against crop failure (Silwana and Lucas, 2002). It is a practice of growing two or more crops on a same field simultaneously, providing mutual benefits to crops. In many small holder intercropping systems are often planted at densities much lower than their usual sole crop density, would still reduce the risk of crop failure (Ibeawuchi, 2007).

Materials and Methods

One intercropping experiment was conducted in farmer's field in Angul District of Odisha state. *Amorphophallus* (Elephant foot Yam) and *Dioscorea* (Grater Yam) were planted in farmer's field as intercrop with Maize, Okra, Papaya, Banana and Arhar in the last week of the month of May. Two lines of Elephant foot yam (EFY) and *Dioscorea* were planted within the space between two rows of banana as well as Papaya. But one line of the Elephant foot yam and *Dioscorea* were planted in the gap of every two lines of maize, okra and arhar. All the crops were harvested at their respective maturity time and the yield was recorded. The net gain was also calculated basing on the market value of the crop. Elephant foot yam and Dioscorea were planted separately with normal recommended spacing of 90cm x 90cm as control. Normal doses of fertilizers were applied to the crops with need-based irrigation.

Nutrient content of Elephant foot yam and *Dioscorea* were calculated. The carbohydrate, protein, dietary fibres and the mineral content in tubers were estimated by standard procedures. Carbohydrate was estimated by the Anthrone method described by Scott and Melvin (1953). Protein was estimated by the procedures described by Lowry *et.al* (1951). Dietary fiber in tubers was also estimated by enzymatic gravimetric method (Prosky *et. al*, 1985). Mineral content was estimated by Rapid spectrometry determination method (Bhandari 2018).

Results and Discussions

The nutritional assessment of EFY indicated that the food value as carbohydrate 70.1%, protein 5.1.%, dietary fiber 3.2% and minerals (3.2%) whereas *Discorea* contained 71.5% carbohydrate, protein 4.5%, fibre 3.0%, minerals 2.8% along with other nutritional supplements (Table 1) by dry weight basis. These tuber crops are very nutritious for consumption, also possess specific medicinal and therapeutic properties (Mishra and Nedunchezhian, 2006; Baah *et.al*, 2009 and Chandrasekar *et.al* 2016). From the ancient times *Amorphpphallus* and *Dioscorea* tubers are used in some indigenous medicinal formulations to recovery health ailments (Mishra and Sriram 2002). Both crops have rich source of carbohydrate and dietary fiber contents might be the definite nutritional supplements for tribal people during food scarcity periods.

Intercropping of elephant foot yam and yam with 5 crops viz. maize, okra, papaya, banana and arhar was conducted in the farmers' field revealed that when elephant foot yam was cultivated as sole crop the tuber yield was 32.52 t/ ha having net gain of Rs. 1,60,000/- per hectare . When in is inter cultured with other crops the production of elephant foot yam was highest 23.26 t/ha with papaya followed by banana (22.81 t/ha). The lowest production was recorded in case of arhar i.e. 16.37 t/ha; 18.43 t/h and 19.2 t/h was harvested when planted with maize and okra respectively. Similarly, when *Dioscorea* was cultivated alone the yield was 25.92 t/ha. The yield was 14.72 t/ha, 14.15 t/ha, 12.93 t/ ha, 12.21 t/ha and 9.26 t/ha when inter cultured with banana, papaya, okra, maize and arhar respectively (Table 2). It was clearly and evidently proposed that intercropping gives higher yields in a given season and greater stability of yields in different seasons compared with sole cropping (Willey, 1979). The highest intercrop yield of yam was observed with banana and papaya having net return of Rs. 1,90,000/- and Rs. 1,02,000/- whereas Rs. 1,08,600/- output benefit reported with arhar because of market value of principal crop. The same net return was recorded in case of EFY intercropped with arhar and papaya (Rs.1,42,000/- and Rs. 1,26,000/-) , where banana resulted the best profitable intercrop with net gain of Rs. 1,90,000/- (added profit over sole principal crop banana).

In intercropping systems yields are more stable, related to yield stability risk factor in terms of either productivity or income or both (Mead and Willey 1980). When EFY intercropped with banana, arhar and papaya as per the market value the net profit was substantially higher than the vegetables. Yam (Dioscorea) also reported net benefit in a same trend as an intercrop. The finding of Nedunchezhian et.al 2002 revealed that EFY increase the net returns (Rs. 21,100/- and Rs. 20,200/-) from banana cultivation without any adverse effect on the main crop. The tuber productivity of greater yam (Dioscorea) was pronounced in association with banana (Nayar and Suja, 2007), also a profitable practice (Nayar and Nair, 1992). Due to their better yield and return, these crops are considered as subsidiary food crops, play vital role in nutrition and livelihood for rural people for employment and income generation (Naskar 2004). Hence, intercropping can increase the income through reduction of economic risk and market fluctuation resulting from growing a single crop which is more prone to climatic hazards and helping the farmers in better utilization of land by having more than one crop produced per unit area (Dodiya, 2018).

Table. 1: Nutrient contents in tubers of EFY and ***Dioscorea***.

Sl.No.	Nutrient components	Nutrient contents (%) in tubers as in dry	
		ETY	***Dioscorea***
1	Carbohydrates	70.10	71.50
2	Protein	5.10	4.50
3	Dietary Fiber	3.20	3.00
4	Minerals	3.20	2.80

Table 2: Yield and Net returns of different crops (Sole crop and Inter cropped)

	Ments / Crops	Yiel Yield of intercrops (t/ha)		Net Return in Rs.		
		EFY	**Dioscorea**	**Principal Crop only**	**Principal Crop + EFY**	**Principal Crop + *Dioscorea***
1	EFY	32.52	--	1,60,000 /--		--
2	Dioscorea	--	25.92	1,30,000 /--		--
3	Maize + Inter Crop	18.43	12.21	27,100/-	61,900/-	56,000/-
4	Okra+Inter Crop	19.21	12.93	28,200/-	66,000/-	58,000/-
5	Papaya + Inter Crop	23.26	14.15	66,000/-	1,26,000/-	1,02,000/-
6	Banana + Inter Crop	22.81	14.72	1,50,600/-	2,35,000/-	1,90,000/-
7	Arhar + Inter Crop	16.37	9.26	72,500/-	1,42,000/-	1,08,600/-

Conclusion

EFY and Yams, energy rich tuber crops served as a source of carbohydrate for larger sections of the people including tribals, although largest portions of tubers produced annually is marketed as the fresh tubers. The rural and food security could be met by adopting local measures to raise the farm output of locally consumed tuber stuffs. Under rain-fed conditions both life support crops are suitable for planting at the last week of May to minimize the irrigation cost. These crops need low production cost and landless tribal farmers also require using forest lands for its cultivation to fulfill their minimum food need during food scarcity periods. However, the qualitative non acrid variety of *Amorphophallus* and high yielding variety of *Dioscorea* are used as commercial crops to uplift their socio-economic status and thus brings food security to the tribal people by providing additional yield from intercropping generating income and employment opportunities thereby enabling sustainable livelihoods in the rural households. Therefore, to achieve food and nutritional securities, increasing household income, the amendments like availability of quality planting materials, technology transfer methods, and improved cultivation practices are required to improve the tribal living status and provide food security along with better livelihood in tribal dominated districts of Odisha.

References

Baah, F.D., Maziya-Dixon, B., Asiedu, R., Oduro, I. and Ellis, W.O.2009. Nutritional and biochemical composition of *D. alata* (*Dioscorea* spp.) Journal of Food, Agriculture & Environment. 7 (2) :373-378.

Bhandari, N. 2018. Technique used in Plant tussue Analysis for essential elements on horticultural plants and correlate with nutrient requirement. North American Academic Resarch. 1(2): 94-113.

Chandrasekara, A. and Kumar, T.J 2016. Roots and Tuber Crops as Functional Foods: A Review on Phytochemical Constituents and Their Potential Health Benefits. International Journal of Food Science Volume 2016, Article ID 3631647, 15pages.

Dodiya, T.P.; Ankit D. Gadhiya and Patel, G.D. 2018. A Review: Effect of Inter Cropping in Horticultural Crops. Int.J.Curr.Microbiol.App.Sci. 7(02): 1512-1520.

Edison, S. 2006. Scope and Strategies for Development of Tuber Crops based Farming system for Food Security and Sustainable Environment. In: Root And Tuber Crops. Eds. S.K.Naskar, K. Rajsekhara Rao, R.C.Ray. & R. S. Mishra. ISBN-81-903825-0-0. pp3-9.

Ibeawuchi II. 2007. Intercropping - A food production strategy for the resource poor farmers. Nat Sci. 5(1):46–59.

Lowry, O.H., N.J. Rosebrough, A. Lewis Farr and Randali, R.J. 1951, "Proteinmeasurement with the Folin Phenol Reagent",J.Biol.Chem.,vol.193,pp.265-275.

Mead, R and R.W. Willey, 1980. The concept of a land equivalent ratio and advantagesin yields from intercropping. Experimental Agriculture, 16:217-228.

Mishra, R.S. and Sriram, S. 2002. Medicinal value and Export Potential of Tropical Tuber Crops, Series Recent Progress in Medicinal Plants, Crop Development, Production Technology and Commerce. Vol. 5, Eds J.N. Govil et. al SCI Tech Pub.USA,pp.376-386.

Mishra, R.S. and Nedunchezhiyan, M. 2006. Commercial cultivation and production of quality planting materials of Yamand Aroids. In: Root and Tuber Crops. Eds. S.K. Naskar, K. Rajsekhara Rao, R.C. Ray. & R. S. Mishra. ISBN-81-903825-0-0. pp 322-330.

Naskar, S.K. 2004. Industrialization or Commercialization of Tuber crops in North eastern regions. In: Pro. Nat. Seminar on Horticulture for Sustainable income and environmental protection. Organised by Dept. of Horticulture, Nagaland University, SASARD, Nagaland, 24-26 Feb. 2004, p84.

Nayar, T. V. R. and Nair, P. G. 1992. Productivity of Amorphophallus paeoniifolius, Dioscorea alata and Xanthosoma sagittifolium in association with nendran banana. J.Root Crops. 18:103- 107.

Nayar T.V.R. and Suja G. 2007. Production potential of root and tubers in association with plantation crops. Proceedings of the 13th ISTRC Symposium, 2007. pp. 186 – 194.

Nedunchezhiyan, M., Jata, S.K., and Sahoo, B. 2009. Intercropping Elephant Foot Yam in Orchard Crops. Orissa Review , October – 2009 p. 82-84.

Prosky ,L., N G Asp, I Furda, J W DeVries, T F Schweizer, B F Harland. Determination of total dietary fiber in foods and food products: collaborative study. J Assoc Off Anal Chem. Jul-Aug1985;68(4):677-9.

Rahman, S.S. 2020. Is elephant foot yam (*Amorphophallus paeoniifolius*) tuber detrimental to patients with diabetes? Diabetes & Metabolic Syndrome: Clinical Research & Reviews.14(5):1015-1016.

Scott, T.A. and Melvin, E.H. 1953. The determination of hexoses with an throne. Analytical Biochemistry, 25:1656-1658.

Silwana TT, Lucas EO. 2002. The effect of planting combinations and weeding and yield of component crops of maize bean and maize pumpkin intercrops. J Agric Sci. 138:193–200.

Willey, R.W. 1979. Intercropping, its importance and research work part-I. Competition and yield advantages. Field crop abstract, 32(1): 1-10.

27

Biofortification Key to Preventing Hidden Hunger

A. Mishra

Introduction

Biofortification is the development of nutrient dense staple food crops using the best conventional breeding practices and modern biotechnology, without sacrificing agronomic performance and important consumer preferred trait (Nestel *et al*., 2006). It provides a comparatively cost-effective, sustainable, and long- term means of delivering more micronutrients (Saltzman *et al*. 2014). This approach not only will lower the number of severely malnourished people who require treatment by complementary interventions but also will help them maintain improved nutritional status. Moreover, biofortification provides a feasible means of reaching malnourished rural populations who may have limited access to commercially marketed fortified foods and supplements. Micronutrient malnutrition (MNM) is the major constrain to provide nutritional security to world population of underdeveloped and developing countries. Among the 20 major risk factors of the global burden of disease estimates, Fe deficiency ranks 9th and Zn deficiency ranks 11th, while in the high mortality countries, including India, they rank 6th and 5th, respectively (WHO 2002). The global significance of MNM known as hidden hunger, came to the attention of the nutrition community as recently as the mid, 1980s. In 1990, The World Summit for Children convoked the governments and facilitated by the UN with support from UNICEF, the World Bank, WHO, FAO, UNDP, CIDA and USAID, was a landmark event in the history of fight against MNM. The development and implementation of biofortification food programs is still in the early stages of growth. Harvest Plus, part of the Consultative Group on International Agriculture Research (CGIAR) Research Program on Agriculture for Nutrition and Health (A4NH), is currently supporting the research and dissemination of staple crops biofortified through conventional as well as molecular plant breeding and through transgenic techniques. Biofortification can be done by both Agronomically and Genetically. Considering the genetic bio fortification process there are several methods like Plant Breeding

Techniques, Molecular Breeding, Mutation Breeding, Genetic Engineering, Tissue Culture etc.

Essential Components to be Fortified Minerals, in the context of the human diet, are inorganic chemical elements (or more properly their dissociated ions) that are required for biological or biochemical processes including the accumulation of electrolytes. There are 16 essential minerals, but 11 of them are required in such small amounts and/or are so abundant in food and drinking water that deficiency arises only in very unusual circumstances. The remaining five are present in limiting amounts in many foods, so a monotonous diet can easily result in deficiency. These minerals are iodine (I), iron (Fe), zinc (Zn), calcium (Ca), and selenium (Se).

Deficiency diseases arise when diets are based predominantly on staple foods, such as milled cereals, which have a low bio-available mineral content (Christou and Twyman 2004).

Vitamin A: Deficiency associated with blindness and increased risk of disease and death for small children and pregnant women can be addressed through supplements, which are now estimated to reach children at least once a year in 40 countries. The UN Standing Committee on Nutrition (UN/SCN) estimates that 140 million children and 7 million pregnant women are Vitamin-A deficient, primarily in Africa and South/Southeast Asia. In 1998, WHO, UNICEF, Canadian International Development Agency, USAID, and the Micronutrient Initiative launched the Vitamin-A Global Initiative. This provides support to countries in delivering Vitamin-A supplements.

Iron: Iron is therefore required for oxygen transport in the body and for energy metabolism, also contributing to the catalytic activity of a range of nonheme enzymes such as ribonuclease reductase (WHO/FAO 1998). The immediate outcome of Fe deficiency is iron deficiency anemia (IDA), which is thought to affect at least two billion people worldwide. More than half of these cases could be addressed by increasing the amount of Fe in the diet, but as for iodine this is difficult in developing countries where the population relies on staples, because cereal grains contain very low levels of Fe and also contain antinutritional compounds such as phytate that inhibit Fe uptake (Zimmermann et al. 2004).

Zinc: Zinc is an essential functional component of thousands of proteins. Many contain zinc prosthetic groups (e.g., zinc finger, zinc twist) and approximately 100 enzymes require Zn as a cofactor. Some olfactory receptors cannot function without Zn. Many cells in the body also secrete Zn as a signaling molecule, including cells in the immune and nervous systems. The mineral appears to be particularly important during periods of rapid growth, and insufficient intake during childhood and adolescence can delay growth, sexual development, and psychomotor development (WHO/FAO 1998).

Calcium: Calcium is the most abundant mineral in the human body, accounting for 1–2 % of an adult's body mass. Over 99 % of Ca is stored in the teeth and bones, where it plays an important structural role (WHO/FAO 1998). Calcium deficiency has a profound impact on bone health, resulting in rickets if deficiency occurs in the young and osteoporosis if it persists into old age.

Selenium : Selenium is found in two unusual amino acids— selenocysteine and selenomethionine—which are the principal functional components of selenoenzymes. It is an essential cofactor in approximately 50 enzymes, including those whose function is to reduce antioxidant enzymes (such as glutathione peroxidase) and those whose function is to remove mineral ions from other proteins (such as thyroid hormone deiodinases). Se is an antioxidant with health benefits including the prevention of cancer and heart disease (WHO/FAO 1998).

Folate: Deficiency associated with increased risk of maternal death and complications in birth and also associated with neural tube defects in infants and with an estimated 200,000 severe birth defects every year can be addressed through fortification of wheat products (Haddad et al. 2 004).

Methods for Attaining Genotypically Biofortified Products Through Plant Breeding

Conventional breeding is limited, however, because it can only use the genetic variability already available and observable in the crop being improved or occasionally in the wild varieties that can cross with the crop. Plant breeding programs focus on improving the level and bioavailability of minerals in staple crops using their natural genetic variation. Breeding approaches include the discovery of genetic variation affecting heritable mineral traits, checking their stability under different conditions, and the feasibility of breeding for increasing mineral content in edible tissues without affecting yields or other quality traits. Breeders utilize molecular biology techniques such as quantitative trait locus (QTL) maps and marker-assisted selection (MAS) to accelerate the identification of high mineral varieties, but they have to take into account differences in soil properties (e.g., pH, organic composition) that may interfere with mineral uptake and accumulation.

Mutation Breeding

This technique makes use of the greater genetic variability that can be created by inducing mutations with chemical treatments or irradiation. The FAO/ International Atomic Energy Agency (IAEA) website contains more than 2500 varieties that have been developed through mutation breeding (Mutant varieties database). Of these, 1568 are in Asia, 695 in Europe, and 165 in the United States. Most of the European and US mutants are flowers, but most in Asia are basic food crops such as wheat, rice, maize, and soybeans. According to their website, FAO/IAEA included bio fortification as one of the objectives of their mutagenesis program, but there do not seem to be any applicable results yet. Varieties produced using mutagenesis can be grown and certified as organic crops in the United States.

Genetic Engineering

Genetic engineering is the latest weapon in the armory against mineral deficiency and uses advanced biotechnology techniques to introduce genes directly into breeding varieties. The genes can come from any source (including animals and microbes) and are designed to achieve one or more of the following goals (Zhu et al. 2007) (a) improve the efficiency with which minerals are mobilized in the soil, (b) reduce the level of anti nutritional compounds, and (c) increase the level of nutritional enhancer compounds.

Tissue Culture

These techniques are now used extensively to produce disease-free planting material of clonally propagated crops such as bananas. When tissue culture is combined with embryo rescue techniques, plant breeders can use the genes from wild and weedy relatives of a crop, which would normally not cross with the cultivated crop. Tissue culture is an important tool for propagation of roots and tubers, such as potatoes and cassava, and both of these crops are part of current bio fortification research.

Some Milestone of Biofortification in Food Crops

1. Rice : Golden Rice-1 (Beta Carotene rich)- Three beta carotene synthesis genes 'psy'(phylotene 6ynthetase), 'lyc' (lycopene cylase) are from Daffodil and one gene crtI (carotene desaturase) from bacterium *Pantoea ananatis pv. uredovora (Erwinia uredovora)*. It contains 5-7 micro gram of carotene per gram of rice.
2. Golden Rice-2 contains 31 microgram carotene per gram of rice. Here the Daffodil gene has been replaced by Maize gene. Other some important

beta carotene rich varieties are IR-64, IR-36, Chehirang, PSBRc 82, BRRI DHAN, SWARNA.

2. Wheat : WB 02(2017)- It contains 42.00ppm of Zn and 40.00ppm of Fe. HPBW 01(2017)- It contains both Zn and Fe as 40.06ppm.
3. Maize : Pusa Vivek HYB.(2017)- Pro Vitamin A(8.15ppm), Lysine (2.67%), Tryptophan (0.74%). Pusa HMY Improved (2017)- Tryptophan(0.91%), Lysine(3.62%).
4. Pearlmillet: HHB 299(2016) contains Fe(73.00ppm) and Zn(41.0ppm). AHB 1200(2016) has Fe (73.00ppm).
5. Mustard : Pusa Mustard 30(2016) has Erucic acid less than 2.0%. PUSA Double Zero Mustard- 31(2016) contains Erucic Acid(<20ppm) and Glucosinolates (<30ppm).
6. Cauliflower : Pusa Beta Kesari- 1(2015) has beta carotene 8.0010.00ppm.
7. Potato: Bhu Sona(2017) contains beta carotene 14.0mg/100gm.
8. Sweet Potato: Bhu Krishna(2017) has anthocyanin 90.00-100 mg.
9. Pomegranate: Solapur Lal (2016) contains Fe (5.6-6.1mg/100g), Zn (0.64-0.69mg/100g), Vitamin C(19.4-19.8mg/100g).
10. Lentil: Pusa Ageti Masoor (2017) has Fe(65.00ppm).

Future Prospectives

1. Expansion on research on pro-biotic and Iron absorption as well as iron fortified crops.
2. Improvement of efficiency of minerals, which are mobilized into the soil and taken up by the plants.
3. Enhancing the storage capacity of the storage tissues which does not participate in plant growth and development rather than available to humans.
4. Reduction of anti-nutritional compound (phytic acid and others) from the edible part of the plants.

Advantages: Increase in nutritional value, reduce micronutrient deficiency disease as well as dietary deficiency disease and provide quick immune response.

Limitations: High production cost of biofortified crops. There is too much chance to get wild varieties by natural crossing of GM crops with others. Low substantial equivalence and does not provide proteins. Bio fortified crop's availability in rural market is impossible. Genetic engineered plants may cause health issues like allergy and nausea.

Conclusion

A sustainable, long term and food-based solution is offered by bio fortification which will positively influence billions of world population in the future that ultimately multiplies the challenges for the researchers or scientists to develop more bio fortified crops. Though, it is cost-effective procedure, for nutritious future biofortification must be taken seriously that will reduce the children and infant mortality ratio. People residing in under developed nations cannot afford the fortified and diverse food. Thus, research and development of more bio fortified crops should be practiced in those areas along with spreading awareness for dietary diversification in order to alleviate micronutrient malnutrition.

References

Christou P, and Twyman RM (2004) The potential of geneti- cally enhanced plants to address food insecurity. Nutr Res Rev, 17:23–42

Haddad L, Ross J, Oshaug A, Torheim LE and Cogill B (2004) 5th report on the world nutrition situation: nutrition for improved development outcomes. United Nations, Standing Committee on Nutrition, Geneva, pp 143. URL: http://www.popline.org/node/236111#sthash.zssi67QS.dpuf

Nestel P, Buois HE and Meenakshi JV. and Pfeiffer W (2006) Biofortification of staple food crops. J. Nutr. 136:1064– 1067

WHO/FAO (1998) Vitamin and mineral requirements in human nutrition: report of a joint FAO/WHO expert consultation, 2nd edn. Bangkok, 21–30 Sept 1998. URL: http://apps.who.int/iris/handle/10665/ 42716#sthash.niRWmJ9o.dpuf

World Health Organization (2001) Water- related diseases: malnutrition. World Health Organization – Water Sanitation and Health (WSH). URL: www.who. int/water_sanitation_health/disease/malnutrition/en/

Welch RM (2002) Comparison of iron bioavailability from 15 rice genotypes: studies using an in vitro digestion/Caco-2 culture model. J. Agric. Food Chem. 50:3586–3591

Zimmermann MB, Wegmueller R, Zeder C, Chaouki N, Biebinger R, Hurrell RF and Windhab E (2004) Triple fortification of salt with microcapsules of iodine, iron, and vitamin A. Am. J. Clin. Nutr. 80(5):1283–1290

Zhu C, Naqvi S, Gomez-Galera S, Pelacho AM, Capell T. and Christou P (2007) Transgenic strategies for the nutritional enhancement of plants. Trends Plant Sci. 12(12):548–555

28

Food and Nutrition Security Challenges and Strategies

G.C. Acharya and M.R. Sahoo

Introduction

India is one among the progressive countries in the world in maintaining food and nutrition database efficiently. The country has been in the fore front in signing the resolution on 'Transforming the World: the agenda for Sustainable Development-2030'. India outstood among the developing countries in food self-sufficiency since seventies with low household hunger rates and low infant mortality rate. In the meanwhile, the country faced double burden of food and nutrition with inadequate dietary intake and under nutrition in one hand and overweight due to low physical activities on the other hand. Rapid increase in population with diminishing food production led to food insufficiency and malnutrition whereas, rapid economic growth and increasing mechanization led to the increasing prevalence of overweight and obesity. Both the under- and over nutrition are associated with health hazards, hence, use of normal body mass index (BMI)would be considered as an indicator for food security (FAO, 2013).

Food Security in Indian Context

India ranks 67 out of 125 countries on the Global Hunger Index (GHI) published by International Food Policy Research Institute (IFPRI). India is home to the world's largest food insecure population, with more than 200 million people who are hungry. The population of India has increased from 361 million (1950-51) to 1250 million (2011). The population is expected to reach 1.44 billion by 2020 and 1.82 billion by 2050. The requirement and shortfall in food grains is expected to be 276 million tonnes and 15-20 million tonnes, respectively by the year 2020 (Niti Aayog, 2018). To realise the projected production of 313.18 million tonnes by 2020, the average food grains productivity has to increase by manifold.

The Challenges Ahead

Population in the country is increasing with diminishing land, water, and other natural resources. India holds 16% of the world population with 2.5% of global land mass. India shares 25% of the global hunger burden with 19 million of undernourished people (WHO, 2018). Four out of ten children in India are malnourished. On the other hands, there has also been an increasing prevalence of obesity among the children in India. India has made significant improvement in rate of under nourishment and malnutrition with low stunting rate among the children. India continues to have one of the world's highest children under nourishment rates which impact the overall productivity in the adolescence. Despite of a large anti-poverty and food security programmes there are critical gaps need to be bridged. The emerging challenges linking to food and nutrient in sufficiency are lower agriculture growth, land degradation, diminishing bio-diversity and climate change.

Strengthening Food and Nutrition Security

United nation defines food security as all the people, at all times, have physical, social, and economic assess to sufficient, safe, and nutritious food that meets their food preferences and dietary needs for an active and healthy life. The term "food security" referred to the ability of a country or region to assure adequate food supply for its current and projected population (Ramachandran, 2013). However, the term "malnutrition" includes both under-nutrition in terms of proteins, calories, fats, vitamins and minerals, and over-nutrition leading to obesity. Food insecurity and malnutrition reasoned due to;

- Over-reliance on few staple food crops
- Low dietary diversity
- Malnutrition in agricultural and food system

About 40% of our daily calories (Fig.1) are met from just three cereal crops viz., rice, wheat and maize out of 30,000 edible plants. Thirty crops fulfill most of the world's calories and nutrient, however, farmers grow only 170 plants on commercial basis. Crop diversification is urgently needed to widen the source of calories from potential high value crops. Closing the production and nutrition gaps requires a transformation of current agriculture and food systems towards greater diversity. Widening food basket diversity is imperative to mitigate food insufficiency. Neglected and underutilized species can be called "hidden treasures" that offer tremendous opportunities for fighting poverty, hunger and malnutrition. Thus, underutilized crops would play an important role in mitigating the malnutrition in the country.

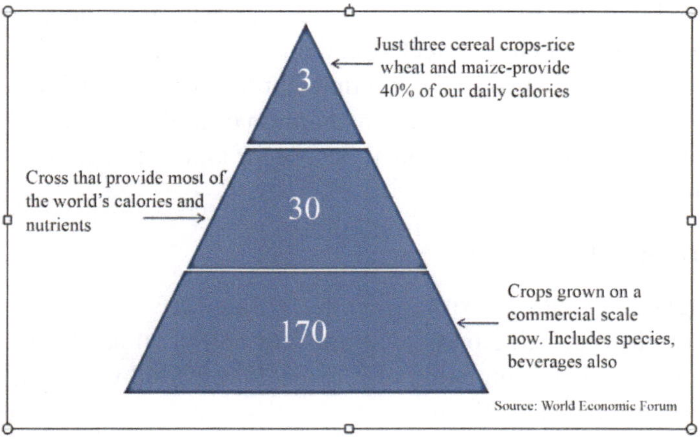

Fig. 1: Dietary pattern for food and nutrient supplements in India

Scientific and technological interventions along with policy issues need to be adopted for improving the agricultural production system which reciprocates the food and nutrition availability (INSA, 2009). The following steps are important to mitigate the hidden hunger in our country;

- Increased agricultural productivity through conventional methods.
- Bio-fortification through conventional breeding methods, molecular breeding and genetic engineering. This includes, Beta carotene/pro-vitamin A–rich sweet potato, and cassava, zinc and iron–rich rice, wheat, maize, pearl millet, and beans; vit. A and iron rich golden rice, protein and amino acid rich transgenic potato using AMA genetic.
- Research-based government supplementary feeding programmes in consultation with nutrition scientists/experts for better impact of the technology.
- Recommended Dietary Allowance (RDA) for important nutrients.
- Nutrition status is to be monitored by the National Nutrition Monitoring Bureau in the entire country instead of selected states.
- Current research on phytochemicals (nutraceuticals) in food to be emphasized.
- Food processing sector to be accelerated to prevent post harvest losses, generate employment, and strengthen nutrition security.
- In addition, there is a need to improve the diversity of Indian diets as reliance primarily on cereals for energy may lead to macro nutrient over-

nutrition coupled with micro-nutrient deficiencies arising from the lack of diversity in the diet which is reflected in the rising levels of obesity. Another factor that contributes to the multiple burdens of malnutrition, is the increased availability and higher consumption of highly processed and packaged food rich in carbohydrates, fats, sugar and salt, increased levels of eating out and snacking.

Vegetables for Nutritional Transition

Of the total 1097 vegetable species grown worldwide, 495 are leafy vegetables, 227 are used for multiple vegetative parts, 204 are root crops, 90 are fruits/ seeds and 81 species are used as comprising of flowers. Of the total vegetable species, we are only familiar with 7% of vegetables that meet out nutrition requirements in our dietary supplements. Promotion of underutilized fruits and vegetable would be an important strategy to overcome the food and nutrition insufficiency,

- To ensure a holistically sustainable development trajectory, the transformation towards vegetable-rich production and food systems should also support climate change adaptation and protection of human and environmental health
- It is highly sensitive to climate change and can enhance vulnerability of producers in water limited areas, as well as exposure to harmful chemical inputs.
- Traditional vegetables are an important asset for meeting this challenge as many have high nutritional value, low water requirements, adaptation to poor quality soils, and good resistance to pests and diseases.
- Several indigenous leafy vegetables (Fig.2) present an optimal source of nutrients such as-carotene, folate, iron, calcium, zinc, proteins and dietary fibre, while showing lower water use and higher water use efficiency in comparison.
- Traditional vegetables can also require relatively fewer labour and economic inputs compared to global vegetables, meaning they present lower risks of financial losses for small farmers
- Tuber crops are the third-most important food source for people after cereals and grain legumes, and boast the versatility to adapt to varying altitudes, climate and edaphic conditions.
- Tuber crops have been a food source since ancient times. Most are grown organically and constitute a major component of indigenous food in tribal areas.

- Orange, yellow and pink-fleshed sweet potato, rich in vitamins A and C, and starch are recognized as a staple food in some African countries. Amaranth including leaf and grain, drumstick, bitter gourd, constitute the most significant portion of biodiversity and are a well-balanced food as they supply vitamins (such as C, A, B1, B6, B9, and E), minerals, dietary fibre and phytochemicals.
- Vegetables also have strong nutraceutical value and can help prevent some chronic diseases and gastrointestinal disorders.
- Most vegetables are short-duration crops that are adaptable to a variety of edaphic and climatic adversities.

Fig. 2: Leafy vegetables

India's Strategies to Zero Hunger

The Zero Hunger Challenge is a global initiative which aims to strengthen the food availability and mitigate hunger. Five indicators have been for Zero Hunger Framework which includes; i. stunting among children below 2 years of age, ii. percentage of undernourished population, iii. percentage depletion of groundwater and percentage of forest cover, iv. productivity among smallholder farmers, income among primary activity workers and self- employed, and, v. percentage of food wasted.

Based on the indicators, India is well maintaining the national food and nutrition databases, undertaking research studies and surveys documenting the ongoing agriculture, food, nutrition and health transitions (Ghosh et al., 2019). Indian scientists have contributed substantially to the global efforts to review the ongoing transitions and evolve appropriate definition of food security, recommendations regarding human nutrient requirements, and develop appropriate standards for assessment of nutritional status (Ramachandran, 2013). The country has also focused on the evidence based intervention programmes to improve food and nutrition security of the citizens, to ensure food sufficiency to meet the energy/nutrients gap among the vulnerable group of people, and to nationalize the prevention, early detection and effective management of under- and overnutrition among the weaker section of the population in a mission mode.

National Food Security Mission (NFSM)

Several missions, programmes and acts such as National nutrition policy (1993), National nutrition plan of action (1995) and National nutrition mission (2001) were implemented in the recent past to strengthen the nutrition goal of India. The missions and policies lag behind due to some unavoidable reasons;

- Nutrition is a poor cousin even in health and agriculture planning and execution.
- Nutrition improvement is not a stated goal with measurable parameters for monitoring.
- Top-down approach without sensitizing the community and making them partners in planning and execution.
- Poor targeting, accountability, and governance.
- Inadequate importance to nutrition in school, college and even professional (health, agriculture, social science) education.
- Neglect of women and children's health and education.
- Vertical programmes with poor convergence and synergy between functioning of ministries and departments.

National food security mission was lunched by the National development council (NDC) in 2007 to enhance the annual production of rice, wheat and pulses by 10.0, 8.0 and 2.0 million tonnes by the end of the eleventh plan (2011-12). The mission was extended with a new target of 13.0 million tonnes of additional food grain production by2019-20.

Policies Interventions for Food Accessibility

In the recent past, India has implemented several successful programmes to ensure food and nutritional security which includes;

- Integrated Child Development Service (ICDS) for the preschool children, pregnant and lactating mothers.
- School Mid-Day-Meal programme (MDM) for ongoing schoolchildren.
- Annapoorna scheme–10 Kg food grains to elderly above 65years.
- National Food Security Act to provide 25 Kg rice or wheat at Rs 3/Kg for families below the poverty line(BPL).
- Public distribution system (PDS) to ensure food availability to a vast segment of undernourished people below and above the poverty line.
- Micronutrient supplementation programmes like i. anaemia prophylaxis programme for distribution of iron folic acid tablets to pregnant and lactating women, children, and adolescent girls, 2.Vitamin A supplement for administration of 100,000 IU of oral vitamin A to 1-6 years old children for me as les immunization.
- Universal iodization of salt to combat iodine deficiency disease
- National Rural Employment Guarantee Act (NREGA).

Conclusion

Agriculture is an essential sector to be addressed for ensuring the food and nutritional security of the country. Indian agriculture confront with the challenge of ensuring food security amidst constraints such as diminishing net cultivated area, deterioration of soil quality, reduced per capita land availability etc. Food insufficiency is thus increasing day by day. Adaptation and mitigation strategies can help minimize negative impacts of food production on Indian agriculture. The need of the hour is, therefore, to synergise modern agriculture research with the indigenous wisdom of the farmers to enhance the food production. This necessitates a strategic approach with a renewed vision and redefined focus to secure our food and nutrition.

References

Ramachandran, P. (2013). Food & nutrition security: Challenges in the new millennium. Indian Journal of Medical Research. 138(3): 373-382.

FAO (2013).The State of Food Insecurity in the World 2013.The multiple dimensions of food security. Rome: FAO.

Ghosh-Jerath, S., Downs, S., Singh, A., Paramanik, S., Goldberg, G. and Fanzo, J. (2019). Innovative matrix for applying a food systems approach for developing interventions to address nutrient deficiencies in indigenous communities in India: A study protocol. BMC Public Health.https://doi.org/10.1186/s12889-019-6963-2

INSA. (2009). Nutrition Security for India–Issues and Way Forward', August 3–4, 2009, New Delhi, India.

Weblinks

https://in.one.un.org/un-priority-areas-in-india/nutrition-and-food-security/https://www.ifpri.org/

http//www.planningcommission.nic.in/plans/planrel/fiveyr/welcome.html https://sustainabledevelopment.un.org/post2015/transformingourworld https://niti.gov.in/sites/default/files/2019-07/WG-Report-issued-for-printing.pdf https://www.who.int/news/item/11-09-2018-global-hunger-continues-to-rise---new-un-report-says

29

Precision Agriculture for Future Food & Livelihood Security

N. Pradhan

The current world population of 7.6 billion is expected to reach 8.6 billion in 2030, 9.8 billion in 2050 and 11.2 billion in 2100, according to United Nations (UN). With roughly 83 million people being added to the world's population every year, the upward trend in population size is expected to continue, even assuming that fertility levels will continue to decline (UN, 2017). This means there will be an extra billion people to feed within the next decade. Continuing population and consumption growth will mean that the global demand for food will increase for at least another 50 years .The future's food systems need to be resource efficient and sustainable. It is quite clear that the only way to grow more food is to increase the yield while saving the environment from further degradation. It is not just food, it is the four F's – Food, Feed, Fiber and Fuel. Digitization has increased in importance for the agricultural sector and is described through concepts like Smart Farming (SF), Precision Farming (PF) and Precision Agriculture (PA).The only way to get higher yield from limited resources is by changing agriculture into what we call Precision Agriculture, Climate Smart Agriculture, or as I prefer to call it Water-Smart Agriculture"

Precision agriculture is comprised of a set of different technologies and methods that combine enhanced types of machinery, information systems, sensors, and informed management decisions that enables the farmer to optimize production by accounting for uncertainties and variables within the agricultural systems. Precision agriculture does not focus only on new technologies; it also focuses on the use of more accurate information fusion and farm management system. Moreover, increased profitability due to development of information systems and better management practices in agriculture.PA techniques/ technologies (PAT) are used by farms or farmers/entrepreneurs in the agricultural sectors of developed and developing countries, in cooperation with various institutions or individually. PA growth rate flattened during the first years of 2000s, because the results (productivity increase, inputs and fuel use reduction, ease of use, low maintenance, and compatibility between brands) were not as positive as expected by the agricultural community. However, PATs are currently taking up again, because technology problems have been gradually resolved and more tangible economic results have been shown at farm level (e.g., profit increase). In 2014, PA global market already amounted to 2.3 billion, with an expected annual growth rate of 12% through 2020 with the mature US and European markets considered the most promising (Balafoutis *et al.*, 2017). Nowadays, PF techniques are used in various countries for both crop and livestock production. The use of PA techniques on arable land is the most widely used and most advanced amongst farmers. PA has the potential to help farmers improve input allocation decisions, thereby lowering production costs or increasing outputs, and, potentially, increasing profits.

However, little is known about how farmers use PATs to support managerial decision-making, or about the relative magnitude of benefits and costs of PATs on individual farms (Batte and Arnholt, 2003). In addition, the factors affecting the adaptation of PATs may vary depending on the characteristics of farms and farmers. Due to the necessity of high-cost technology, PATs started in developed countries and the adaptation problems of farmers in these countries were encountered. As can be seen from the examples in the literature, it is seen that mostly cotton farmers have adapted to the methods which can be used effectively in large-scale farms. The use of PATs in the production of basic foodstuffs also shows improvements in some countries. It is recognized that PATs can play an important role in the European Union(EU) in meeting the growing demand for food, feed and raw materials, while ensuring the sustainable use of natural resources and the environment. However, the adoption of the PA in the EU also faces certain challenges due to the size and diversity of farm structures (EU, 2014). Factors influencing the adaptation of PATs can also be considered as factors limiting the development of these techniques. Therefore, some facilitating measures and some support systems need to be developed by decision-makers and governments to ensure the expected benefits from these agricultural techniques. PATs have a real potential to deliver a more productive and sustainable agricultural production, based on a more precise and resource-efficient approach. However, while in the USA possibly up to 80% of farmers use some kind of PATs, in Europe it is no more than 24% (Smart-akis, 2018). Moreover, PATs are used for both crop and animal production activities, in large agricultural holdings and in small family businesses and even in urban agriculture. However, there have been some factors limiting the development and spread of the PATs, these factors include cultural perception, lack of local technical expertise, infrastructure and institutional constraints, knowledge and technical gaps and high start-up costs with in some cases a risk of insufficient return on the investment (EU, 2014)

The use of technology in the agricultural sector has been increased over time. Modern technology is used to improve the wide types of production practices employed by farmers that make the works of the farmers easier by increasing productivity, profit and quality in agriculture have become smarter. Taking advantage of these developments for sustainability food security is among the priorities of almost every country. Now, both developed and developing countries are also adopting this technology. The factors needed for the developments of PATs are considered as high technology cost, data security, lack of information and institutional cooperation within the limits of this study. It is thought that decreasing the irrigation water use, especially in agricultural production all over the world, will motivate every country in the near future in order to meet the current needs related to PATs.

References

Balafoutis, A., B. Beck, S. Fountas , J. Vangeyte, T. der Wal , I. Soto , M. Gómez-Barbero, A.Barnes , and V. Eory, 2017. Precision Agriculture Technologies Positively Contributing to GHG Emissions Mitigation, Farm Productivity and Economics, Sustainability 2017, 9, 1339.

Batte, M. T.and M. W. Arnholt, 2003. Precision Farming Adoption and Use in Ohio: Case Studies of Six Leading-edge Adopters, Computers and Electronics in Agriculture 38 (2003) 125 /139.

EU, (European Union), 2014. Precision Agriculture: An opportunity for EU Farmers -Potential Support with the CAP 2014-2020, Directorate-General For Internal Policies, Policy Department B: Structural and Cohesion Policies, Agriculture and Rural Development.

Smart-akis, 2018. https://www.smart-akis.com/index.php/network/what-is-smart-farming/. Accessed: 01 October 2018.

UN (United Nations), 2017. Department of Economic and Social Affairs. https://www.un.org/development/ desa/en/news/population/ world-population-prospects -2017. html. Accessed: 02 October 2018.

References

30

Nutritional Security *vs.* Food Security In India

S.S. Nanda, S. K. Mohanty, S. Samantaray and S. Swain

Nutritional and food security is the problem of the day globally due to increase in population by many folds. It is therefore one of the important challenges of public and private sector to provide nutrient rich food with appropriate security for development of human kingdom. Advances in food production have largely kept pace with demand. More than 830 billion people are still hungry and more than 2 billion people are lack of sufficient nutrients. Paradoxically, more than 2 billion people are overweight or obese. Malnutrition means bad nutrition not only under nutrition. For some time the food debate centered on the food security drawing on wildly accepted thinking from 1966 i.e. world food summit which defined food security as a state or condition "when all people at all times are physical, economic and social access to sufficient, safe and nutritious food to meet the dietary needs and food preference for active and healthy life". Thus nutrients are necessary component to food security..

In 2012, the world food security set the bench mark as "when all people at all times have physical, social and economic access to food which is safe and consuming in sufficient quantity and quality to meet their dietary requirement. Food preferences must be supported by an environment of adequate sanitation, health services and care, allowing for a healthy and active life". More recently concern has been expressed on malnutrition, overweight and obesity. This is due to over consumption of energy dense food, insufficient exercise and genetics. Together, it leads diet related diseases such as type-2 diabetes and cardio vascular diseases. We should be clear about the words like nutrients, nutrition and nutritious. Nutrient is a constituent, nutritious is a descriptor or state of condition. We should think about nutritional security rather than food security. Nutritional security encompasses food security which further encompasses nutrient content.

India is in serious stage with respect to hunger index with the score of 31.3% which is more than South Asian countries like Sri Lanka, Bangladesh and

Nepal. People are vulnerable to access food. This vulnerability has created great space between food and people. This space can be filled up by ensuring food availability, accessibility and affordability. Though, there is an increase trend in the intake of pulses, oils, fats and vegetables by the children in the country, there exits 30% gap in the actual and recommended dietary energy intake of the people, though the per capita income has increased almost 1400 in 2016. The latest phytonutrient report indicated that younger generations consume less nutritional food than elder persons. The recent report by NITI Ayoga and UN report on December 18 reveal that Himachal Pradesh, Kerala and Tamilnadu have gained sufficient score to achieve no poverty and no inequality. Tamilnadu has scored 68 as the state has taken various supportive measures for poverty eradication, for good health and wellbeing.

India has produced good quality of food and vegetables with annual production of 0.88 and 162 tonnes respectively. Despite India being the second largest producers of fruits and vegetables in the world, India's consumption is less than recommended quantity of 400g per person per day. The poor intake of fruits and vegetables is largely determined by dietary choice of individuals which is skewed towards cereals, low fruits and vegetables intake are estimated to cause 31% heart diseases and 11% stroke worldwide. In addition there was decline in milk and milk products and overall calorie consumption in children below age of 3 years. Low intake nutrient rich food affects children's future productivity negatively. The dietary pattern and intake level of foods among children of tribal of Odisha, Chhattisgarh, and Bihar is poor. The level of malnutrition in tribal areas of India is high, which is due to less number of breast feeding, delayed initiation of supplementary nutrition and implementation of poor growth and poor malnutrition reduction interventions.

Malnutrition problem and its associated diseases can be reduced only when the people follow the recommended dietary pattern including adequate intake of fruits and vegetables. A diet high in fruit and vegetable is associated with decreased risk of serious chronic diseases like cancer and cardio vascular diseases. It is estimated that 2.7 million lives can be saved each year if fruits and vegetable consumption is increased.

Adequate amount of millets could protect diabetes as the crop contains high level of protein, minerals, vitamins, antioxidants, non-acid forming, low glycaemic index, non-starch polysaccharides and dietary fibres. Livestock products such as fish and milk rich in protein possess richer content of amino acids, unsaturated fatty acids, vitamins and trace metal. It is easily digestible. Consumption of fish controls blood pressure, cardiac diseases, cholesterol and cancer in human beings. The other side of malnutrition is over nutrition.

Southern states like Tamilnadu and Andhra Pradesh suffer from obesity or over weight.

Hunger, malnutrition and obesity have become a global challenge. This affect the sustainable development of nation in multiple aspect. The national institute of nutrition recommends an individual should take 300g. vegetable per day, including 50g. green leafy vegetable, 200g.other vegetables, 50g. of tubers apart from 100g. of foods per day. This type of food during 0-5 years help to strengthen the physical and knowledge level, more importantly overcome malnutrition problems which is major challenging area of sustainable development goal.

Even there is no research is undertaken to analyze the dietary diversity pattern among children in India as children are most vulnerable to poor dietary diversity which causes malnutrition, micronutrient deficiency and it is associated disease and disorders such as scurvy, anaemia, blindness, diarrhoea, respiratory illness, infectious diseases consequently problem of growth retardation and poor school performance.

31

Entrepreneurship Opportunities in Cultivation of Edible Mushrooms in Odisha

K. B. Mohapatra

Mushrooms is known for its delicious and nutritious food with highly digestible lysine rich protein, vitamins and minerals, Mushroom serves as a low calorie food which lacks fats and are low in carbohydrate. They are rich in folic acid, phosphorus, potassium, calcium, copper and iron and vitamin-B complex. In place of starch, mushroom contains sorbitol and linolenic acid. They are excellent sources of thiamine, riboflavin, niacin, pantothenic acid, biotin, folic acid and vitamin B-12. Further, mushrooms are also reported to have high medicinal attributes. A large number of mushrooms contain biological active polysaccharide protein complex having anti-tumor, immune-modulating and antioxidant properties.

People of Odisha have accepted mushroom as a food item since ancient time. A wide range of edible mushrooms grow in the forests during the rainy seasons. Commercial mushroom cultivation was initiated in the state in1969-70 by the than Prof. G. C. Rath, a renowned fungal taxonomist in the Department of Plant Pathology, OUAT, Bhubaneswar. Under his leadership, the Government of Odisha started a mushroom production unit for cultivation of paddy straw mushroom which became popular within a short time and many entrepreneurs came up to start the commercial cultivation of paddy straw mushroom by 1970. Later Odisha University of Agriculture and Technology established a research and extension center at Bhubaneswar campus in the name of Center of Tropical Mushroom Research and Training (CTMRT) to commercialize the cultivation of paddy straw mushroom (*Volvariella vovacia*), oyster mushroom (*Pleuratus sajorcajo*) and milky mushroom (*Clocybe indica*) as the principal crops. Mushroom cultivation offers scope to recycle agro-wastes as carbon pool into good quality protein, much of which is otherwise wasted in the field. Mushroom cultivation could be an effective component in the Integrated

Farming System (IFS) as practiced in the state of Odisha. At present the state produces 19,603 tons of mushrooms per annum contributing to over 12 per cent of the country's total production

Integrated Farming System and Mushroom Cultivation

Unlike most of the development efforts, in mushroom cultivation, the emphasis has been on economic development with improvement in quality of life of the rural poor through an integrated effort. Labor intensive nature of cultivation, significant value addition, dependence on locally available abundant and cheap raw material, low initial investment, simple technology and limited space requirement make it a worthwhile proposition. Mushroom, being an indoor crop, happens to be women friendly in nature. This brings about removal of poverty and social development in harmony and therefore, fits well in any farming system. Integrating mushroom cultivation in existing farming systems will supplement the income of rural poor, provide gainful employment and will lead to inclusive growth.

Global Scenario of Mushroom Production

China is the largest producer, consumer and exporter of mushrooms in the world followed by USA and Netherlands. China grows over 40 mushroom species in a small to industrial scale. Ten most widely cultivated mushrooms in China include *Pleurotus ostreatus, Lentinus edodes, Auricularia auricula, Agaricus bisporus, Flammulina velutipes, A. polytricha, P. cornucopiae, Coprinus comatus, Agrocybe chaxinggu* and *Volvariella volvacea* (Li, 2012). World mushroom production has gradually increased from 0.30 million tons to 34 million tons over a period of last 55 years from 1965 to2020. The three major mushroom producing countries as per FAO data namely China, USA and Netherland account for more than 60 % of the world production. However, share of China itself is 46 % which is about half of the world mushroom production. According to current Indian estimates, mushroom production of India is about 1, 55,553 metric tons with an average growth rate of 4.3%, which is 0.5 % of the world mushroom production. The button mushroom commands the largest share of 73 per cent of the total production followed by oyster (16%), paddy straw (7%), milky (3%) and others (1%).

Mushroom production in India started in the 70s with phenomenal growth rate, both in terms of productivity as well as production. In seventies and eighties button mushroom was grown as a seasonal crop in hills, but with the development of technologies for environmental controls and increased understanding of the cropping systems, mushroom production shot up from mere 5000 tons in 1990 to over 1,00,000 tons in 2010. Today, commercially grown species are button and oyster mushrooms, followed by other tropical mushrooms like paddy straw mushroom, milky mushroom etc. Two to three crops of button mushroom are grown seasonally in temperate regions with minor adjustments of temperature in the growing rooms; while one crop of button mushroom is raised in North Western plains of India seasonally. Oyster, paddy straw and milky mushrooms are grown seasonally in the tropical/ subtropical areas from March to October. The areas where these mushrooms are popularly grown are Odisha, Maharashtra, Tamil Nadu, Kerala, Andhra Pradesh, Karnataka and North Eastern region of India.

Scope of Mushroom Cultivation

- Waste biomass of paddy is suitably be utilized for commercial mushroom production at a low cost.
- It can be grown both as a indoor crop and outdoor crop under the shade of fruit orchards.
- The cultivation technology is women friendly.
- No problem in selling of fresh mushroom, particularly the paddy straw mushroom as the market demand is high.
- Mushroom production centers in a locality with technical knowledge for production of quality mother culture and spawn which can be sold to small growers.
- Value added products like, prickles, jam, jelly, dry mushroom powder for soup making which can be linked to the market chain.
- The partially decomposed mushroom straw can be suitably converted in to high quality compost.
- Mushroom can be cultivated all round the year by growing different mushrooms coinciding suitable climatic conditions.

Mushroom Cultivation in Odisha

Paddy Straw Mushroom (*Volvariella* spp.)

The paddy straw mushrooms are mostly preferred by the people for its flavor and nutritive value. The only disadvantage of the mushroom is its low self life. Straw mushroom is now being cultivated in costal districts of Odisha largely as an intercrop in the coconut plantations (Fig. 2) and under the sheds of other perennial plants (Fig. 1) (Thakur and Mohapatra, 2013). It is always advised to cultivate straw mushroom under protected thatched roof to avoid intervention of insects, reptiles, the toads and other stray animals. Farmers are mostly cultivating *V. volvacea* as a secondary source of income, making use of their waste paddy straw of their won field. Mostly, cultivation of straw mushroom is done using non-pasteurized substrate with incorporation of pulse powder or wheat bran in order to provide instant nutrition for quick establishment of the mushroom mycelium. It is more prone to competitor moulds and diseases resulting low biological efficiency. However, there is scope for improvement in biological efficiency by using more productive strains of the species, producing good quality spawn and effective substrate management. Further, higher and more stable yield (30-45 %) could be obtained through adoption of indoor method of farming (Ahlawat and Tewari, 2007, Reyes and Abella, 1997) using partially composted substrate. The state of Odisha produces about 3.05 million tones of paddy straw per annum and there is enough scope for mushroom production at a low investment.

Fig. 1: Commercial cultivation of paddy straw mushroom under shed of coconut plantation

Fig. 2: Outdoor cultivation of paddy straw mushroom under bamboo plantation

For low cost cultivation, good quality paddy straw bundles are collected, made to 45 cm length by trimming both the ends and soaked in 2 % solution of calcium carbonate for six hours. Bundles are taken out, excess water drained off to 65% substrate moisture and beds are raised having dimension of 1.5x1.5x1.5 (length x breadth x height). Depending on the straw length the size of the bed may be increased to minimize the wastage of straw. Requirement of spawn and organic additive has been standardized at 3 % each of the dry weight of the substrate. Three layered raised beds are prepared for spawning and organic supplementation at 1:1:2 proportions. Beds are topped with a thin layer of straw followed by covering with polythene sheet that maintains appropriate temperature for mycelial ramification (30-35°C) and fruit body formation (28-30°C). Polythene sheet is removed after the emergence of pin heads; watering is done as and when required and the fruits of the first flush are harvested at 12-15 days (Fig. 3). The second harvest from the same bed is done at 21-22 days provided proper moisture level of the bed is maintained. One bed thus prepared, may yield approximately one kilogram of fruits with a biological efficiency of 15 % with appropriate aftercare.

Fig. 3: Paddy straw mushroom

Milky mushroom (*Calocybe indica*)

About two decades ago, milky mushroom was identified as a wild edible mushroom in India. However, complete commercial production techniques were developed for the first time in Tamil Nadu (Krishnamurthy, 1995). Milky mushrooms are attractive with bright white coloration. Initially it look like button mushroom which later become much larger in size. Milky mushrooms are tropical mushrooms like *Volvariella* spp. but having higher level of productive and self life with biological efficacy of 80-100%. It can be grown successfully on substrates like paddy, wheat or barley straw and other different types of cellulosic agricultural residues. Besides the substrates, 100 grams of quality seed spawn and 10 to 15 grams of any one of the pasteurized maize mill, wheat bran, paddy husk or boiled wheat are required per one bag of the growing substrate.

For cultivation of milky mushroom the paddy straws are chopped to 4-5 cm size and soaked in clean water for 6 hours. The excess water is drained out to the level of 50-60% of moisture. One third of the soaked straw along with the pasteurized substrates is placed in polythene bag of 60 X 40 cm size to a depth of 7-8 cm and pressed uniformly. One third of the seed spawn is spread along the periphery of the layer. Likewise, layering and spawning is repeated three times. Holes on the polythene bags need to be made both on the side and bottom of the poly bag for proper ventilation. The inoculated bags are incubated in a dark room. Room temperature of 20-25°C and 80% relative humidity is ideal for growth and production of milky mushroom. It will take about 20 days for colonization of the substrate. Well sterilized casing soil (50:50 soil and compost mixture) may be spread on the top of the mycelia growth which provides physical strength to the fruiting bodies of the mushroom to grow erect. The mushroom buds will be ready for harvest within a week of their appearance or when the size of the fruiting body reach 7-10 cm in diameter.

Oyster mushroom *(Pleurotus* spp.)

Oyster mushroom has species suitable for both temperate and sub-tropical regions. It is the third largest cultivated mushroom of the world. Odisha produces 7,025 tons/annum during the winter months (November-February) contributing to 46 per cent of total oyster mushroom production of the country. The mushroom can be cultivated in a wide range of temperature (22-28°C) employing most cellulosic farm wastes. The production technology of oyster mushroom is almost same as the milky mushroom except that no covering with casing soil is required. Further, the poly bag is removed after appearance of the fruiting bodies (Fig. 4). It is a fact that though the cultivation of oyster mushroom is feasible during the winter season but not appreciated much by the consumers for fresh consumption in Odisha in comparison to straw mushroom. However, this could be a very good raw material for prickle making.

Fig. 4: Oyester mushroom at fruiting stage

Button mushroom (*Agaricus bisporus*)

The button mushroom is most popular variety of mushroom in the country. At global level, it ranks first in terms of production. Punjab is the leading state contributing to 60 per cent of the total production of the country. Being a temperate mushroom, production can be taken up seasonally during winter months in vast hilly tracts of the interior districts of Odisha during September to March. Odisha has just entered into commercial production with 214 tons / annum and it is likely to grow further in near future.

Economics of Mushroom Cultivation

Mushroom cultivation is a profitable enterprise. The cost involved in raising one bed of straw mushroomof1.5x1.5x1.5size comes toRs.60/-with a production of one-kilogram mushroom within a crop cycle of 21 days. The net return is Rs.60/- per bed assuming the market rate at Rs.120/- per kilogram. Likewise, the cost for raising one bag of oyster mushroom is Rs.40/- per bag with a production of 1.5-kilogram mushroom within a crop cycle of 45 days. The net return is Rs.50/- per bag assuming the market rate at Rs.60/- per kilogram. A model small mushroom production unit (300 sq. ft.) with the investment of Rs.25,000/- accommodating 120 beds of paddy straw mushroom per month during summer and rainy season and 225 bags of oyster mushroom per two months during winter season gives an estimated net income of Rs.10,000/- per month.

Table 1: Mushroom production in Odisha (tones/annum)

Sl.No.	Mushroom	Production (tons/year)	%of total production
1.	Paddy straw	12,364	63
2	Oyster	7,025	36
3	Milky	0	0
4	Button	214	1
	Total	**19,603**	**100**

Table 2: Economics of the production system of a bamboo-based structure (25'x12')

Sl.No.	Item	Quantity	Rate (Rs.)	Amount (Rs.)
1	Bamboo	40 Nos.	150/-	6,000/-
2	U.V. proof polythene sheet	80 Meters.	200/-	16,000/-
3	Fogger/Micro sprinkler	One set	8,000/-	8,000/-
4	Exhaust fan	One	2,500/-	2,500/-
5	Consumables	-	2,500/-	2,500/-
6	Labor wages	20 Nos.	250/-	5,000/-
7	Total	-	-	40,000/-

Protected (Indoor) Cultivation of Mushroom

During the winter season (November-February) under Odisha condition, outdoor cultivation of paddy straw mushroom is not recommended only because of prevalence of low temperature and very low humidity. Hence, considering the resource base of the mushroom growers of the state an attempt was made by the Precision Farming Development Centre (PFDC) in collaboration with All India Coordinated Research Project on Mushroom to grow paddy straw mushroom in the low cost bamboo based structure designed and promoted to maintain appropriate temperature and humidity for cultivation of paddy straw mushroom during the winter season. The economics of the production system under the referred structure is given below:

It is imperative to say that mushrooms cannot be grown year after year with full commercial access, unless proper growing conditions are provided and adequate facilities are available for the control of competitor moulds, diseases and insect pests. Possibly, such conditions can be fulfilled in shelf growing, by the construction of properly insulated and ventilated mushroom houses accommodating store room, spawn running room, cropping room as well as packing and preservation room.

Keeping in view of the higher sale price of the produce, off season cultivation or winter cultivation of paddy straw mushroom is gaining popularity now-a-days in the state. Hence, poly house cultivation of paddy straw mushroom is being popularized during winter season wherever growers are interested. It is possible to maintain required temperature (34-35°C) and humidity (80 %) in the poly house through provision of foggers or micro sprinklers and therefore, satisfactory yields can be obtained from the tropical mushrooms even during winter months. Low cost poly houses with 200-micron ultraviolet proof polythene sheets erected even on bamboo frame can be raised by the growers for the purpose. Length wise, on the rear wall, a vent with an exhaust fan has to be provided for evacuation of hot air. Poly mushroom growing houses can be of variable dimensions. However, the appropriate dimension for maintaining suitable microclimate and obtaining good yields could be 20 x10 x9 (length x breadth x height).

Indoor Cultivation using Partially Composted Substrate

Indoor cultivation of paddy straw mushroom with partially composted substrate can be a boon for mushroom growers in view of its higher productivity. This gives 30-45 % yield in comparison to 10% obtained from outdoor farming. Semi-industrialization of paddy straw mushroom cultivation can be made possible utilizing this technology. The substrate with paddy straw and cotton waste in equal proportions each partially composted for a period of six days with addition of calcium carbonate and rice bran in appropriate amounts. The compost is spread on shelves in the growing room and steam pasteurized at 60°C for six hours. After natural cooling, spawning is done and left covered with polythene sheet. After the spawn run is over, light and intermittent fresh air is introduced into the growing room for induction of fruit bodies. Suitable temperature and humidity are maintained inside the growing room. A quantity of 30-45 kg of fresh mushrooms can be harvested in two flushes within the crop cycle of 21 days per each 100 kg of compost. Hence, this system is far more productive than the outdoor/indoor cultivation. This system has successfully been initiated in Odisha and efforts are on to popularize it among the farmers.

Spawn Production

In nature, mushroom spores help in survival of aspecies from one generation to the next. These spores are extremely small and therefore, difficult to handle as seed. Further, spores need time and specific conditions to germinate and during this time the competitor fungi grow faster in comparison to mushroom fungus. Therefore, a pure culture of the desired mushroom mycelium is first raised on a convenient artificial culture medium and then added to a substrate to give it an advantage. Spawn is a vegetative mycelium from a selected mushroom grown on a convenient medium. It comprises mycelium of the mushroom and a supporting medium, which provides nutrition to the fungus during its growth. Spawn is used as seed in mushroom cultivation. Appropriate kind and quality of spawn is very important for mushroom cultivation. The isolation, purification and maintenance of mushroom cultures require technical expertise and aseptic laboratory facilities. It is difficult to maintain pure cultures and/or spawn for small mushroom growers. They have to rely entirely on commercial spawn producers. Government and non-government organizations that play a vital role in supplying quality spawn of a desired mushroom strain or variety. At present Odisha is ahead of other states in establishment of commercial spawn production centers. It has been estimated that a quantity of over 4047 tones of spawn is produced per annum from 266 spawn production units present across the state in both private and public sectors. However, mushroom productivity often declines owing to use of poor-quality spawn supplied by some spawn producers. In this context, it is imperative to produce quality spawn in order to exploit the yield potential of different mushroom species.

Maintenance of Spawn Quality

A major source of contamination for the growing mushroom mycelium is the grain seed or the substrate used to prepare the spawn. Modern equipment and facilities for sterilization and the maintenance of sterile conditions, however, are capable of reducing fungal and bacterial contamination to 0.1% or less of the number of spawn units. Quality control in spawn making consists of constant inspection to eliminate those units visibly contaminated or those units exhibiting unacceptable differences in appearance and growth. Spawn of most mushrooms can be refrigerated, but it should be kept in room temperature until the temperature of the spawn is brought down to normal before it is used as inoculums or as planting spawn. *Volvariella* spawn, in particular, should not be kept either at refrigerated or room temperatures for more than four weeks after mycelia growth are complete. The presence of pink chlamydospores in *Volvariella* spawn indicates a highly fertile and active growth, although their time of appearance and density depends upon the strain and substrate used.

Recycling of Spent Mushroom Substrate (SMS)

Natural plant wastes are the basic substrates for production of all species of edible mushrooms, capable of utilizing lignin, cellulose and hemicelluloses and producing fruit bodies with excellent nutritional and medicinal attributes (Rajarathnam *et al.*, 1997). Mushroom growing is an eco-friendly activity as it utilizes the wastes from agriculture, animal husbandry, poultry, brewery etc. But dumping of SMS after crop harvest may cause serious environmental problems including ground water contamination (Beyer, 1996). Compost is considered as 'spent', when one full crop of mushroom has been taken and further extension of cropping become non-remunerative. Spent mushroom compost being rich in organic matter adds nutrients to the soil, helps in neutralizing acidic soils, facilitate plant growth in barren areas and also it improves water quality along with bioremediation of contaminated soils. However, for improvement of the physico-chemical characteristics of SMS for its use as manure, re- composting methods are employed.

The paddy straw mushroom spent substrate has pH in the range of 8.82 – 9.16; while oyster mushroom spent substrate has pH within 6.51-7.69. The oyster mushroom contains higher nitrogen (1.82%) as compared to paddy straw mushroom (1.06-1.46%). The other nutrients like organic carbon and phosphate are higher in paddy straw mushroom spent substrate than in oyster mushroom spent substrate. However, contents of potassium, magnesium and calcium are at par in two types of SMS. The addition of SMS in nutrient poor soil improves its health by improving the texture, water holding capacity and nutrient status. Incorporation of SMS leads to an increase in both pH and organic carbon content. Experiments have shown that, the dry matter content of plants increases with the incorporation of increasing amount of weathered or non -weathered SMS in soil (Chang, 1982). The mixing of spent mushroom substrate in soil has shown plant growth promoting activity in several plant species.

Constraints Encountered in Mushroom Production

Production and supply of quality spawn holds the key in order to improve production and productivity. A large quantity of spawn produced in the state is being sold to the farmers of the neighboring states for mushroom production. Further, there is no mechanism to ascertain the spawn quality of mushrooms like other crops which often deteriorates productivity. There is urgent need to establish at least one model spawn production unit in each block of the state for narrowing down the demand supply gap. Moreover, both straw and oyster mushrooms are being grown as intercrops in coconut plantations leading to large scale incidence of contaminants, diseases and insect pests, bringing down the productivity to sub-optimal level.

Marketing of fresh mushrooms is not being undertaken in an organized manner. Mushrooms being highly perishable commodities, suffer a lot owing to marketing in polythene pouches in warm and humid conditions. As production is increasing day by day, possibility of glut cannot be ruled out, which focus on the establishment of more and more processing units in the state. It will stabilize market price and reduce wastage of the produce.

Majority of our entrepreneurs are resource poor and they are not accessible to loans extended by financing agencies. Diversification of mushroom species is minimal in the state as well as the country. Growing of 12-15 crops of straw mushroom in a particular area year-round brings down productivity by the gradual build up and spread of contaminants, diseases and insect pests. Owing to the limitations in pesticide application in mushrooms, they are difficult to manage. Therefore, it is imperative to diversify the species/varieties for sustainable production and productivity. Myths prevailing in the minds of the people like the non-veg status of mushrooms have to be cleared through mass awareness programs.

Conclusion

The geographical area of the state of Odisha is divided into four major agro-climatic zones starting from high altitude hilly regions to coastal plains which are most suitable for growing of all different types of mushrooms. Traditionally the tribal used to collect different naturally growing edible mushrooms from the vast forest tracts of the state for their consumption. In the coastal plains the most preferred edible mushroom is the paddy straw mushroom. Therefore, the commercial cultivation of straw mushroom has been widely accepted by the farmers both in urban and rural areas. Low biological efficiency and vary low self life are the major constraints which need for the attentions of the researchers, growers and the policymakers to make it a more profitable entrepreneurship venture.

References

Ahlawat, O.P. and Tewari, R.P. 2007. Cultivation technology of paddy straw mushroom (Volvariella volvacea). Technical Bulletin, National Research Centre for Mushroom, Chambaghat, Solan (H.P.). p.36.

Beyer, M.1996. The Impact of the Mushroom Industry on the Environment. Mushroom News 44(11):6- 13.

Chang, S.T.1982. Cultivation of Volvariella mushrooms in South- East Asia. In: Tropical Mushrooms: Their Biological Nature and Cultivation Methods (Chang and Quimio Ed.). The Chinese University Press, Hongkong.493 p.

Krishnamoorthy, A.S.1995. Studies on the cultivation of milk mushroom, Calocybe indica P. and C., Ph.D. thesis, Tamil Nadu Agricultural University, Coimbatore, India. P.124.

Rajarathnam, Shashirekha, M.N., Zakia Bano and Ghosh, P.K.1997. Renewable Ligno-cellulosic wastes, the growth substances for mushroom production, National Strategy. In: Advances in Mushroom Biology and Production (R.D. Rai, B.L. Dhar and R.N. Verma Ed.), Mushroom Society of India, NRCM, Solan, India. Pp.291-304.

Reyes, R.G. and Abella, E.A. 1997. Mycelia and basidiocarp performance of Pleurotus sajor-caju on the mushroom spent of Volvariella volvacea. Proceedings of International Seminar on the 'Development of Agri-business and its impact On Agricultural Production in South-East Asia'. Tokyo NODAI press.Pp.491-497.

Thakur, M.P. and Mohapatra, K.B. 2013. Tropical Mushrooms: Present Status, Constraints and Success Story. Proceedings of Indian Mushroom Conference, 16-17 April, 2013. PAU, Ludhiana (Punjab), Pp.42-43.

32

Study on Marketing of Cotton in Rayagada District of Odisha

M. Mishra and S. Mishra

Introduction

Cotton is a fiber crop of Malvaceous family. There are four cultivated species of cotton viz. *Gossypium arboreum, G. herbaceum, G. hirsutum* and *G. barbadense* out of which *G. hirsutum* is the predominant species contributing about 90% to the global production. In India all the above four cultivated species are grown on commercial scale as a kharif crop. India is the largest producer as well as consumer of cotton in the world. The major cotton growing states of India are Maharashtra, Punjab, Haryana, Rajasthan, Uttar Pradesh, Madhya Pradesh, Gujarat and parts of Andhra Pradesh & Karnataka. In Odisha cotton cultivation is an up-coming activity. Many of the western belt farmers, mainly from Rayagada, Bolangir, Kalahandi, Nabarangpur, and Nuapara districts have been taking cultivation of this cash crop. According to the report of Odisha economic survey, 2018-19 a total of 1.45 lakh ha area was cultivated in 2017-18 producing 3.65lakh bales of cotton. In Rayagada district cotton is cultivated in 9 blocks out of 11 blocks comprising an area of 35170 ha. For the year 2018-19 yield was 12.37 q/ha with production of 25.6 MT of cotton according to the data given by District agriculture office, Rayagada. Rayagada being an underdeveloped district lack an efficient marketing system which is very much essential for farmers to realize their return. Further, the farmers of the area are facing transportation and communication problems causing a hindrance in marketing of cotton. Against this back drop, the present study was carried out to calculate the price spread and marketing efficiency of cotton in different marketing channels.

Material and Methods

The study was carried out in Rayagada district of Odisha. Rayagada is one of the major producers of cotton in Odisha so it was selected purposively for the sake of convenience for data collection. Rayagada block was selected randomly and a total of 120 farmers i.e. 40 farmers from small, 40 farmers from medium and 40 farmers large size groups were selected at random from five randomly selected villages. A list of village level traders was also prepared with the help of sample farmers from the selected five villages and from this list 10 village level traders were selected randomly. A pre tested questionnaire was used to collect primary data from cotton growers and traders through personal interview.

Analytical Procedure Used

Total marketing cost: It refers to the sum total of the cost incurred by the producer and various intermediaries in transferring the produce from producer to consumer i.e. labour charge, weighing charge, storage charge, loading and unloading charge etc. It was calculated using the formula,

$$C = Cf + Cm1 + Cm2 + \dots\dots + Cmi$$

Where,

C = Total marketing cost

Cf = Cost paid by the producer from the time the produce leaves the farm till he sells it.

Cmi= Cost incurred by i^{th} middleman in the process of buying and selling the product.

Price Spread

It refers to the difference between the price paid by the consumer and the price received by the producer. It is calculated using the formula,

Price spread = Consumer's price – price received by farmer Ps = Cp – Pf

Where, Ps = Price Spread, Cp = Consumer's price, Pf = Price received by farmer

Marketing Margin

Marketing margin refers to the difference between the price paid and received by any specific marketing agency. The general expression for estimating the margin for intermediaries is as follows,

Intermediaries Margin = Gross/ sale Price – Price paid – Cost of marketing– Loss in the value

Marketing efficiency

The marketing efficiency has been calculated by using the modified method as suggested by Acharya and Agarwal (2001),

MME = RP/ (MC+MM)

Where, MME = Modified measure of marketing efficiency RP = Price paid by consumer or retailers sale price, MC = Total marketing cost and MM = Net marketing margin

Result and Discussion

In cotton cultivation, cost of cultivation incurred is high so to have a good return and profit, producers require a good marketing system with efficient marketing channels. Marketing channels show how products pass from producers to consumers through different marketing agencies or intermediaries.

Table 1: Channel wise quantity of cotton sold (in q)

Sl.No.	Marketing Channel	Size of group			
		Small	Medim	Large	Total
1	I	15.5 (89.7)	28.2 (83.93)	39.6 (76.04)	83.3 (80.91)
2	II	1.78 (10.3)	5.4 (16.07)	8.2 (15.75)	15.38 (14.94)
3	III	-	-	4.28 (8.22)	4.28 (4.16)
4	Total	17.28 (100)	33.6 (100)	52.08 (100)	102.96 (100)

(Figure in parentheses show percentage to the total)

In district Rayagada there were mainly three marketing channels of cotton which were,

Channel I: Producer- Village Merchant- Ginner- Consumer

Channel II: Producer- CCI- Consumer

Channel III: Producer- Ginner- Consumer

The detailed channel wise information on the quantity of produce sold through different channels by the cotton growers is presented in Table 1. From the table it can be inferred that Channel I was the most preferred channel among producers as 80.91 percent of total produce was sold through this channel. In Channel II, marketing through procurement agency was done. In Rayagada the procurement agency is Cotton Corporation of India which procured 14.94 percent of total produce on an overall level. It can be seen that only large farmers who has resources for transportation sold their produce directly to ginners. Through Channel III, only 4.16 percent of total produce was sold. Ginners after ginning of seed cotton sold their produce to consumers. In case of CCI, they had their own cotton processing facility in which, after ginning of cotton they sold to potential buyers.

In case of small farmers, they sold most of their produce i.e. 89.7 percent of total through Channel I and 10.3 percent through Channel II. They didn't sell any produce directly to ginner due to lack of transportation facility. Similarly, medium size group also sold most of their produce i.e. 83.93 percent of total through Channel I and remaining 16.07 percent through channel II. Large farmers also sold major portion of their produce through Channel I i.e. 76.04 percent of total but they sold less i.e. 15.75 percent to CCI, instead they preferred to sell the remaining to ginner directly for more profit.

Various costs are incurred for performing various marketing operations like labour cost, weighing cost, storage cost, transportation cost, ginning charge, loading and unloading charge etc. which together results in marketing cost. Price spread refers to the difference between total price paid by the consumer and total price received by the producer. Marketing efficiency shows how efficiently produce are marketed through different channels. In table 2 marketing cost, marketing margin of various marketing intermediaries price spread and marketing efficiency of different channels are given.

From the table it can be observed that return received by producer was highest in Channel III i.e. Rs. 5700 followed by Rs. 5500 in channel II and Rs. 5000 in Channel I respectively even though price paid by consumer was same in all the three channels. Total of marketing cost and marketing margin was highest in Channel I i.e. Rs 2520 followed by Rs.2060 and Rs.1800 in Channel II and Channel III respectively.

Table 2: Marketing cost, marketing margin, Price spread and marketing efficiency of different channels (Rs/q)

Sl.No.	Particulars	Channel- I	Channel- II	Channel- III
1	**Producer**			
i	Gross price received by the farmer	5000 (69.4)	5500 (76.4)	5700 (79.17)
ii	Marketing cost	320 (4.4)	360 (5)	380 (5.28)
iii	Net price realized	4680 (65)	5140 (71.4)	5320 (74)
2	**Village Merchant/ CCI**			
i	Price paid	5000 (69.4)	5500 (76.4)	-
ii	Marketing Cost	132.42 (1.84)	660 (9.2)	-
iii	Marketing margin	667.58 (9.27)	1040 (14.4)	-
iv	Price received	5800 (80.56)	7200	-
3	**Ginner**			
i	Price paid	5800 (80.56)	-	5700 (79.2)

Sl.No.	Particulars	Channel- I	Channel- II	Channel- III
ii	Marketing Cost	540 (7.5)	-	540 (7.5)
iii	Marketing margin	860 (12)	-	960 (13)
iv	Price received	7200	-	7200
4	**Consumer**			
i	Price paid	7200 (100)	7200 (100)	7200 (100)
5	**Price spread**	2200	1700	1500
6	**MC+MM**	2520	2060	1880
7	**Marketing efficiency**	2.86	3.5	3.83

(Figure in parentheses are percent to price paid by consumers)

It can be observed from the Table 2 that price spread decreased and price received by producer increased as number of intermediaries decreased in the channels. Price spread was highest in Channel I i.e. Rs.2200 followed by Rs.1700 and Rs.1500 in Channel-II and Channel - III respectively. Marketing efficiency also followed a reverse trend. It was highest in Channel-III i.e. 3.83 followed by 3.5 and 2.86 in Channel II and Channel I respectively. It can be inferred that as the number of intermediaries decreased, marketing cost and marketing margin also decreased leading to increase in marketing efficiency.

Conclusion

Channel I was the most preferred channel where 80.91 percent of total produce was sold followed by Channel II where 14.94 percent of total produce was sold. Channel III was preferred by only large farmers due to availability of resources for direct communication with the ginner. Return received by farmer was highest in Channel III which was Rs.5700 and price spread was highest in Channel I which was Rs.2200 indicating as the number of intermediaries increased return received by the farmers decreased and price spread increased. Marketing efficiency was highest in Channel III (3.83) followed by Channel II (3.5) and Channel I (2.86) indicating as the length of marketing channel increased marketing efficiencydecreased.

References

Activity report of Department of Agriculture and Farmers' Empowerment, 2017-18

Rani P and Gupta S. 2017 Marketing Channels, Marketing Margins, Costs and Price Spreads: A Case Study of Bathinda District of Punjab, IRA International Journal of Management and Social Sciences, 7(2):294-301.

Singh H and Sidhu MS. 2007. Marketing of cotton seed in Punjab, Agricultural Marketing. 50(1):15-20.

Verma AR, Rajput AM and Kashyap LR. 2002 The Economics of Production and Marketing of cotton in Khargone district of Madhya Pradesh, Indian Journal of Agricultural Marketing, 16(2):18-27.

33

Agricultural Finance Issues and Challenges

N. N. Mishra

Introduction

Agriculture is an important sector of Indian economy and it contributes about 17.4% to the country's Gross Value Added (GVA) for the year 2016-17 provides employment to over 54.6 % of the population. There has been a continuous decline in the share of agriculture and allied sector in GVA from 18.6% in 2013-14 to 17.4% in 2016-17. Falling share of agriculture and allied sector in GVA is an expected outcome in a fast growing and structurally changing economy. However since 54.6% of the population is still dependent on agriculture and engaged in improving the agricultural production and overall income. The agricultural production system required timely and adequate agriculture credit to fulfill farmers need. Agricultural Finance ensures provision of multiple types of services dedicated to support both on- and off-farm agricultural activities and businesses including input provision, production, and distribution, wholesale, processing and marketing.

Situation of Agriculture in Finance

In order to understand the progress of agricultural credit and its performance with respect to agricultural GDP, the ratio of agricultural credit to agricultural GDP is computed taking into account the agricultural credit outstanding as well as disbursement. The contribution of agricultural credit to agricultural GDP in the country since independence is mentioned in the table below.

Source of Agricultural Finance

NAFIS Report 2016-17 has revealed that agricultural households avail credit from sources other than banks such as NBFC/MFI, financial companies, financial corporations, provident fund, insurance, relatives, friends, moneylenders, landlords, etc. The chart below shows the distribution of agricultural households according to type and sources of loans According to NAFIS, institutional sources were preferred by agricultural households to avail credit as approximately 61 per cent of them avail credit from them. However, a significant portion, i.e. approximately 30 per cent of agricultural households still avail credit from non- institutional sources, which is a cause of concern. The probable reasons could be that their credit demand could be for consumption purposes or they could be tenant farmers, sharecroppers and landless labourers who are not able to offer collateral security to avail institutional credit. They are involved in unviable subsistence agriculture or banks don't find them credit worthy. As a result, these farmers find it convenient to borrow money from non- institutional sources due to easy accessibility.

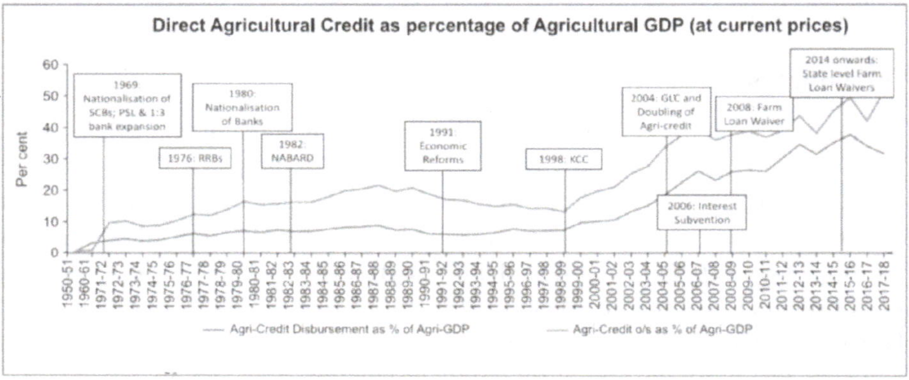

Fig. 1:

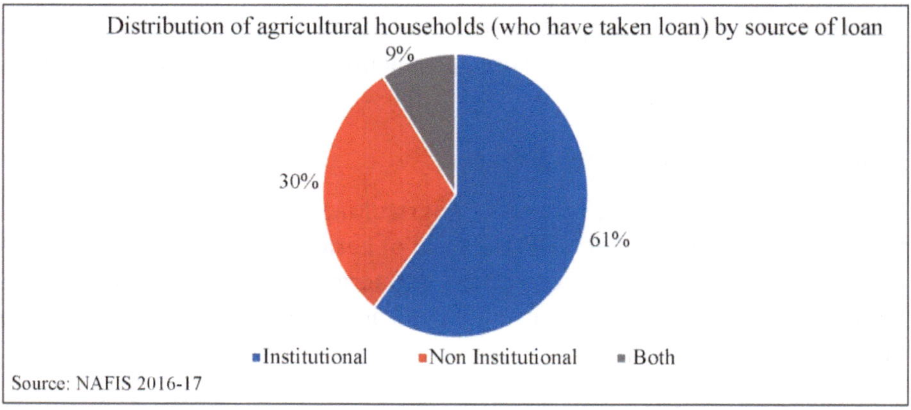

Fig. 2:

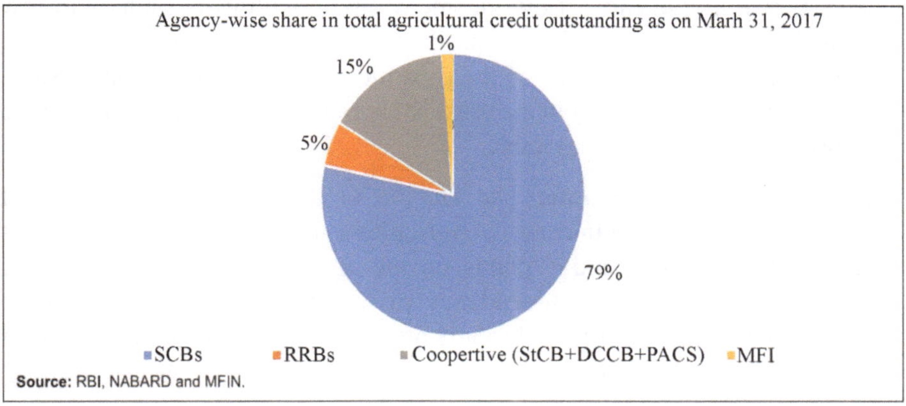

Fig. 3:

Institutional Agricultural Credit - Agency-wise share

In order to understand the performance of different agencies in lending to agriculture sector, a comparative analysis has been done showing their share in total agricultural & allied credit outstanding as on March 31, 2017.

Scheduled commercial banks contributed the major share (78 – 80 per cent) in agricultural and allied credit. Co-operative institutions also play a significant role in extending agricultural credit and the share of all co-operative banks/ institutions constituted 15-16 per cent. The RRBs contributed the remaining 5 per cent of the agricultural credit.

Need of Credit in Agriculture

No sector in the economy can develop without the availability of timely and adequate finance. Agriculture is no exception to this. Since Indian agriculture experiences low returns and uncertainties due to its dependency on natural resources. As most of the farmers India are small and marginal farmers with land holding less than two hectares, the income remains low and they are inclined towards borrowing. The role of farm credit in strengthening and developing both input and output markets in agriculture is crucial and significant.

The following factors are responsible for the need of credit in Indian Agriculture:

1. **Low Income of the farmers:** The income of an agriculturist is very much lower than that of his counterpart in non-agricultural activity. As per the Budget 2018, the average income of a farmer is estimated at Rs 77,976 per year (Rs 213 per day). This is very low as compared to the

cost of living in India. Low income of an agriculturist results into low savings, low investment, low productivity and the farmer continues to remain in the lower strata of the pyramid of the Indian economy. Since the farmer has to invest in the seasonal agricultural operations, hence the borrowing becomes inevitable.

2. **Uncertainty in Income:** The income of the farmer in India is not only low but also uncertain. Agriculture in India often experiences natural calamities and depends on the vagaries of monsoon. Taking into consideration the dependency of the farmers on these uncertain environmental conditions, farmers are prone to opt for funds in case of crop failure or low income.

3. **Increase in Expenditure:** The cost of cultivation has been increasing due to adoption of modern techniques of cultivation which are highly capital intensive. Modern inputs like improved seeds, fertilizers, pesticides, tractors, harvesters etc. are costly and to purchase these inputs, farmers have to borrow.

4. **Seasonality:** Agriculture being a seasonal activity, farmer cannot start farming when he has funds, but has to spend in various agricultural activities during the particular season. Since agriculture is a biological activity, it is not possible to reduce the gestation period. One cannot get the output before the usual duration of time. Requirement of credit arises largely due to a seasonal production cycle coupled with consumption expenditure spread throughout the year.

5. **Supporting Agri allied businesses:** In order to realize enhanced income,, an Indian farmer has to carry out allied businesses activities like poultry, dairy, fishery, and other petty businesses along with the normal crop production activities. The farmers require funds to run these enterprises in business mode.

6. **Diversification and commercialization of agriculture**: In order to maximize the profit, the farmer has to adopt modern technologies and farming methods requires higher amount of funds.

Challenges in Farm Finance

Faced by Farmers

1. Problems on adequate and timely availability of credit
2. High interest rates

3. Procedural delays
4. Inconsistent income (due to price fluctuation)
5. Staggered land holding resulting in lower funding
6. Lack of awareness on the government schemes

Faced by Lenders

1. Inadequate credit outreach due to low infrastructure and manpower
2. Higher transaction cost
3. Risky profile as majority are dependent on rainfed agriculture.
4. Poor recoveries
5. Political interventions (Loan Waivers)
6. Sarfaesi (Securitization and reconstruction of Financial Assets & Enforcement of Security Interest Act) is not applicable on Agricultural land.

Ways to Enhance the Outreach of Farm Credit

Agricultural credit is just one of the tools to modernize farming and needs to be coupled with improved farming practices and policy measures. This can help the farmer to realize farm income which in turn will improve repayment capacity. The important measures to boost farm credit are discussed below.

- **Growing of less water intensive crops:** Since there is scarcity of rainfall and water availability in many regions in India, there is a need to identify and grow the less water intensive crops. Crops like millets which require lesser water can be grown in areas which have scarcity of water. 18 percent of the world's population which resides in India only has access to 4 percent of usable water sources. This would lead to lesser dependence on monsoon and lesser uncertainty of income for farmers. The sustained income is capable of repaying the loan and eligibility for fresh lone.
- **Crop diversification:** is one of the options to enhance the productivity. Many parts of India, over ages are growing the same crop (called as Monoculture). This result in depletion of soil fertility, lower crop production, increased infestation of pests and diseases and hence resulting in a lower crop yield. Farmers can opt for intercropping or growing of new crops basis the water availability, soil fertility and

climatic condition of the region. The farmers shale take the benefit of crop loan through scale of finance fixed for different crops.

- **Use of Hi-Tech agri farming** can increase the quality and quantity of yield. Use of green houses, polyhouses can be effective in areas where there is a frequent climate change and less water availability. Crops can be grown in controlled environmental conditions which can improve the productivity. Use of Drip irrigation, Sprinkler irrigation in the areas with less water availability has proven a good option for optimum utilisation of available water resources. Farmers can have access to loans meant for High-tech agricultural farming.
- **Agri allied activities:** The repayment capacity of the farmer increases when there is a supporting income by the way of agri allied activities like Poultry, Dairy, Fish farming, etc. The income generated through these activities act as buffer for the farmer in case of any adverse unforeseen situation. The growth of agri allied sectors is more than the Agriculture growth in India. E.g. India's dairy industry is expected to maintain 15 % compounded annual growth (CAGR) over 2016-20 with 14.8% average annual growth rate, India stood first in the world in the export of fisheries products. Farmers can easily diversified loans for agriculture and allied activities as per their requirements.
- **Warehousing and storage facilities:** Farmer ends up selling his produce at low rate than the expected price during off season. India's grain production has steadily increased due to advances in technology, but post-harvest loss is constant at 10%. Losses during storage, account for around 6% of the total losses as proper storage facilities are not available. Proper, sufficient and scientific storage facilities can help farmers store their produce after harvest and sell the produce at a time when the market price is high. The multipurpose cold storage facility can helpers the farmers to store the perishable agri produce like fruits and vegetables which will fetch high price when market is favorable.
- **Contract farming** has proved to be a better earning model for the farmers. Farming contracts are agreements between the farmers and the buyers (usually a corporate which processes and markets the end produce) that stipulate what the farmers will grow and how much they will grow with financial and technical support and guaranteed purchase by the business firms.
- **A Producer Company** is formed by 10 or more farmers dealing in agricultural produces or post-harvest processing activities. A Producer Company (Farmer Producer Organization) is an association having an

objective of production, harvesting, procurement, grading, pooling, handling, marketing, selling, export of primary produce of the members (farmers). A group of farmers come together and decide on the different agricultural activities to be undertaken to ensure better income through economies of scale. They have better bargaining power through this model. The banks provide a lenient lending support.

- Creation of special Food Processing zones: While the food processing industry is a large industry, it is still at a nascent stage in terms of development. Out of country's total agricultural and food produce, only 2% is processed. The highest share of processed food is in the Dairy sector, where 37% of the total produce is processed, of which only 15% is processed in organized sector. Creation of these special food processing zones at different regions can help farmers get a better price for their produce and can reduce the cost of storage and transportation. Processing of Fruits and vegetables (Raw Fruits and vegetables, Fruit pulp extraction, Canned fruits and pickles, etc) , Meat and Poultry processing , Packed food units (Noodles/ Vermicelli) , Fisheries (Marine Fisheries, Frozen products and Minced Fish products), staple foods (Flour mills, sugar processing, etc), beverages (Fruits based drinks, carbonated drinks) are some of the processing activities that can run in these special zones. The farmer entrepreneurs have easy access to bank credit to carry out their activities in special food processing zones.

References

Current Issues in Agriculture Credit in India: An assessment – Reserve bank of India Occasional Papers Vol. 28, No. 1, summer2007.

Economic Survey 2016-17. Government of India. Ministry of Finance. Department of Economic Affairs. Economic Division. January,2017

Krishi Annual Report 2016-17. Department of Agriculture, Cooperation and farmer welfare. Ministry of Agriculture and Farmer Welfare. Government of India, Krishi Bhawan.

NABARD Annual Report2016-17

Report of the Internal Working Group to Review Agricultural Credit, September 2019. Issued by Reserve Bank of India.

Role of Rural Credit in Indian Agriculture – International Journal of Advance Research in Science and engineering. Vol No. 5, issue No 03, March 2016.

34

Constraints Faced by Rural Women Entrepreneurs in Dairy Sector

S.J. Jena, I.P. Pattnaik and A.P. Kanungo

Introduction

Women entrepreneurship in economic development of any nation has been recognised for its significant contribution. The women folk can be considered as backbone of any nation. Now time has come for women empowerment. In rural Odisha, women not only play an important role for agricultural development but also plays significant role in allied agricultural sector such as dairy, fishery, poultry, mushroom production etc. In Indian condition dairy has been considered as "Rural Banking" as it is practiced mostly by majority of the marginal, small and landless farm families for their sustenance and livelihood. The role of rural women for animal husbandry is of paramount importance. In this sector they are managers as well as decision makers. Women do more than 80% dairy operation like cleaning, feeding, milking, marketing of milk and milk products. The state and central government initiate several scheme, programmes and projects for empowerment of women through various micro enterprises including dairy. Against this background a study was undertaken to identify the constraint faced by women entrepreneur in dairy sector in Khordha district of Odisha.

Research Methodology

A study was conducted in the Khordha district of Odisha state. Both purposive and random sampling techniques were followed for selection of blocks, panchayats, villages and respondents. The sample size was 120 women entrepreneurs of animal husbandry sector. The response was obtained from each individual respondent through a pre tested structured interview schedule on the major constraints faced by the women entrepreneurs such as family related constraints, technical constraints, marketing related constraints, financial constraints, socio-personal constraints. The response was collected accordingly in a 3 point continuum i.e Strongly Agree, Agree, Disagree with a scale value of 3, 2, 1 respectively.

Result and Discussion

The aim of the study was to identify the constraint faced by the women entrepreneur in dairy sector under five major areas of constraint, such as family related constraints, technical constraints, marketing related constraints, financial constraints, socio-personal constraints. Entrepreneurs generally face different kinds of problems and this problem assume more seriousness in case of women entrepreneur as they face gender discrimination by solving them. Among family related constraint lack of support from family members, multiple roles of conflicts, strains in managing home were the major constraint. In order of importance which ranked as second, third and fourth respectively. Under technical constraints in adequate training skills, lack of suitable technology, non-availability of animal health care facility, were major constraint which ranked first, second, third in order of importance. Under marketing related constraint lack of proper marketing facility, low price of milk, lack of preservation facility, risk orientation towards processing and other value adding activities, lack of knowledge of diversification of product were the major constraint in order of importance first, second, third, fourth, fifth respectively. Under finance constraint lack of awareness of different government schemes, subsidies and their procedures for availing difficulty in getting credit, high rate of interest perceived as major constraint in order of severity and ranked first, second, third respectively. A major problem experienced by women entrepreneur under socio-personal constraint, such as stress due to dual role followed by risk and uncertainties were the major problems which act as a barrier for women entrepreneur in dairy sector. The findings of Bhagyalaxmi (2002) and Usha Kiran (2012) were in the same line of the present finding of constraint faced by women entrepreneurs in dairy sectors.

Table 1: Constraints faced by women entrepreneurs (N=120)

S.No.	Statement	Frequency	Percentage	Rank
A.	**Family related constraints**			
1.	Lack of support from family members	90	75	I
2.	Multiple role conflict	26	21.66	II
3	Strain in managing home	4	3.33	III
B.	**Technical constraints**			
1.	Non- availability of animal health care	9	7.5	III
2.	Lack of suitable technology	18	15.0	II
3.	Inadequate training skills	93	77.5	I
C	**Marketing constraints**			
1.	Lack of knowledge of diversification of product	4	3.33	V

S.No.	Statement	Frequency	Percentage	Rank
2.	Lack of proper marketing facilities	58	48.33	I
3.	Lack of preservation facilities	20	16.67	III
4.	Low price of milk	22	18.33	II
5.	Risk orientation towards processing and other value adding activities	16	13.33	IV
D.	**Financial constraints**			
1.	High rate of interest	18	15	IV
2.	Lack of awareness of different government schemes	60	50	I
3.	Difficulty in getting credit	20	16.66	III
4.	Subsidies and their procedures for availing loan	22	18.33	II
E.	**Socio-personal constraints**			
1.	Time management	15	12.5	II
2.	Stress due to dual responsibility	93	77.5	I
3.	Risk and uncertainties	12	10.0	III

Conclusion

Irespective of the problems and challenges that emerge in the process of entrepreneurship, women are going ahead in the business domain. The findings of the study will be helpful to administrators, policy makers and extension functionaries of the government, to solve the problems of women empowerment in animal husbandry sector and to take remedial measures to attract more number of rural women to promote entrepreneurial activities in dairy sectors.

References

Bhagyalaxmi .C.(2002). A Critical study on micro-enterprises management among women entrepreneur of Andhra Pradesh. Unpub, Phd Thesis, ANGRAU, Hyderabad

Vani.CH.S (2013). Constraint in dairy micro enterprises faced by women entrepreneur of Andhra Pradesh. Indian Res. J.Ext.Edu, 13 (3), pp-106-108

Usha Kiran V. (2012). A Study on business women in micro enterprises. International. Journal of Multi disciplinaryResearch.2(1),66-7

35

Entrepreneurship in Agriculture and Allied Sectors

A. K. Behera

Introduction

Agriculture and allied sectors have always been the backbone of the Indian economy and despite concerted industrialization in the last six decades agriculture still occupies a place of pride. The significance of agriculture in the national economy arises from the role it plays in India's national income, employment generation and export earnings; this can be seen from the provisional estimate on contribution of various sectors to the GDP in the first quarter of 2020-21, particularly during COVID-19 pandemic, positive percentage change (+3.4) over previous year when all other sectors have negative change.

India as a country with huge natural resources has several inherent strength for growth of agriculture and allied sector. It been bestowed with varied agro-climatic zones, which are suitable for raising different crops and livestock species. The demand for agri-inputs like quality seed & planting materials, fertilizer (both organic and inorganic), pesticides, agro-machinery/ equipment, feed and fodder is continuously growing. Availability of raw material is huge, can be used as the necessary inputs for getting a final product. There is ample scope for exports for earning valuable foreign exchange, will strengthen national economy. There is a strong traditional knowledge base, percolated from the generations to generations, which is beneficial in manufacturing like artisans industry, textile industry, cashew industry, handicraft industry etc. Vast population of the country creates large domestic demand. All of these provide vast scope for entrepreneurship in agriculture and allied sector.

Entrepreneurship in Agriculture and Allied Sector

Enterprises carrying out all activities those are performed from Farm to Plate such as supply of all kind of agricultural inputs and production and transformation of agricultural products/produces, and their distribution to final consumers can be established and managed successfully by entrepreneurs. Agriculture and Allied Sector have been explained hereunder for better understanding.

Agriculture: All field crops such as cereals, pulses, oilseeds etc. fall under agriculture.

Horticulture & Allied sectors: This includes fruits and vegetables, flowers, plantation crops, spices, aromatic & medicinal plants, Mushroom, Apiary etc.

Sericulture: It is also known as 'Silk Farming'. It is the rearing of silk worms for the production of raw silk. India is ranked as the second major raw silk producer in the world and it's famous for its silk products worldwide. It is one of the most labor intensive sectors, combining activities of both agriculture (sericulture) andindustry.

Animal Husbandry Dairying and Livestock: Dairy farming for milk production, Livestock rearing particularly small ruminants such as goat, pig etc. are under this category.

Fisheries and Aquaculture Sector (Fresh water, Marine): The 'fisheries and aquaculture sector' is recognized as the sunshine sector in Indian agriculture. It stimulates growth of number of subsidiary industries and is the source of livelihood for a large section of economically backward population, especially fishermen. This covers both capture and culture including inland and sea, aquaculture, gears, navigation, oceanography, aquarium management, breeding, processing, export and import of seafood, special products and byproducts.

Opportunities

A list (indicative but not exhaustive) of possible areas of entrepreneurship is indicated below.

- Application of Bio-technology in production of seed, bio-control agents, industrial microbes for bakery products.
- Export possibilities with respect to Cereals, Spices and Condiments, Fruits & Vegetables, Flowers, Medicinal plants and Essential oils, Meat, Milk and Milk products, Fish and Fish products, Ornamental Fish, Forest by products etc.
- Production of Vegetables and Flowers under controlled environment.
- Secondary and tertiary processing of agricultural commodities.

- Vast coastline and internal water courses (Marine Fish, Inland Fish and Ornamental Fish)
- Livestock rearing.
- Bee Keeping and apiary
- Mushroom production and processing for domestic and export market
- Organic farming
- Vermi-compost production
- Integrated/ composit e-farming
- Production of Bio-pesticides and Bio-control agents.
- Marketing including e-commerce.
- Infrastructure Development (Transport, Cold Storage, Warehousing etc.)
- Logistic support.
- Agriservices

However, we have to keep it in mind that customers in the new normal are subject to more scrutiny in terms of food hygiene, food safety, food traceability etc.

Challenges

- Fragmented hand holdings (Large number of small/ marginal holdings)
- Low per capita income of farmers (Low investment - LowProductivity)
- Poor access to technology and credit.
- Complexity of geographical spread, diversity of crops and large number of farmers' involvement.
- Multiple intermediaries
- Disconnected supply chain (Disrupted now due to COVID-19pandemic)
- Inadequate infrastructure
- Poor access to market and market information
- Seasonal availabilities

Threats

- **Global Competition:** Globalization has led to a competitive business environment in terms of price, quality and service. The agri-entrepreneur, to be successful has to face such competition from rising presence of large corporate/ multinational companies in the market place.
- **Unorganized market:** In India agriculture marketing is too much unorganized.
- **Bad trade practices:**Absence of fare trade practices like compromising in quality of product, weight, packaging etc. pose serious problem to this industry.
- **Price fluctuations:** Due to variation in demand supply gap price fluctuations doesn't permit an enterprise to adopt a long term sustainable pricing strategy
- **High cost of modern technology:** The micro, small and medium enterprises fail often to access modern technologies due to prohibitive high cost.
- **Political biasness**: Policies are often biased in favour of large enterprises.

Government Support

- Specific measures granted to address the "burden of debt servicing" (Agricultural term and crop loans) due to COVID19 pandemic.
- APMC Act is being revised.
- Cluster Development, formation of FPOs and Farmer Producer Companies are being encouraged.
- Provision of Capital Investment Subsidies for establishment of commercial agri- enterprises, processing industries etc. has been made under various Central/State Govt. Schemes.
- Special initiative for infrastructure development and fisheries Development announced recently.
- Promotional organisations like APEDA, MPEDA, Spices Board, Coffee Board &Tea Board, Coconut Board and Cashew Directorate.

36

Climate Change Impact and Soil Temperature: An Overview

J.M.L. Gulati and N.K. Dhal

Agricultural production system has gone a long way in supporting the world's food requirement. The projected estimates revealed that to meet the global population of about 11.3 billion in 2050, the global demand for cereal will increase by 60% (Rosegrant and Cline, 2003, Bengtsson *et al.*, 2006). Agriculture is considered as the chief agent of environmental transformation. The unrestrained attitude of Homo sapiens towards the use of natural resource catalyzing the causes leading to global warming, has increased the challenges manifold to provide food security in a sustainable manner. Increased concentration of CO_2 and other green house gases have resulted in an increase of 0.76^0 C in surface temperature globally since nineteenth century and the mean global surface temperature is predicted to increase by an additional 1.3^0C to 1.8^0C by 2050(IPCC, 2007). Sinha and Swaminathan (1991) reported that an increase of 2^0C in temperature could decrease the rice yield by about 0.75 t/ha in the high yield areas and a 0.5oC increase in winter temperature would reduce wheat yield 0.45 t/ha. Whereas Rao and Sinha (1994) reported that wheat yield could decrease between 28-68 % without considering the CO_2 fertilization effect, and would range between 4 to -34% after considering the CO_2 fertilization effect.

Soil is a three phase system comprising of solid, liquid and gaseous phases. The complex nature of soil is the outcome of interplay of these and also influenced by diversified practices that farmers raise during the crop production process. Soil is exposed to radiant energy which the difference between insolation of short wave length and net outgoing long wave radiation. How much heat flux into the soil?, is determined by soil thermal properties. Change in atmospheric temperature will also have a profound effect on soil thermal regime. Climate change impact thus, bears a correlation with the soil temperature. Krishnan (1992) stated that temperature of the earth surface responds more quickly than the air temperature to change in weather. In the first meter of the soil there are rapid changes in temperature and lapse rate in association with weather changes, similar to those occurring in the first km. of the atmosphere. The effect of large scale weather irregularities and seasonal changes would penetrate to much lower depth.

Effect of soil temperature on crop growth: Plants undergo two different types of environment during its growth; below the soil and above the soil. Thus, variations in either side affect on the following growth stages of plants.

Seed germination and seedling emergence (Table-1): Seed germination process depends upon imbibitions of moisture for enzymatic and respiratory activities. The differential temperature requirement for crop varies due to physiological differences. Chaudhary and Ghildayal (1969) reported favorable temperature range for germination to be 26.5^0 to 37.5^{0}C in wheat while rice seedlings emerged only in the temperature range of 15 to 35^{0}C. Bierhuizen (1973) provided the cardinal temperature for some of the important crops.

Table 1: Cardinal temperature for germination of some field crops.

Crops	Cardinal temperature (^{0}C)		
	Minimum	**Optimum**	**Maximum**
Wheat	3-4.5	25	30-32
Barley	3-4.5	20	38-40
Maize	8-10	32-35	40-44
Rice	10-12	30-32	36-38
Tobacco	13-14	28	35
Sugarbeet	4-5	25	28-30
Peas	1-2	30	35
Oat	4-5	25	40
Sorghum	8-10	32-35	40
Lentil	4-5	30	36

Shoot growth and grain yield: Both extremely high and low temperatures are detrimental to shoot growth. Bierhuizen (1973) found that yield of corn almost increased linearly as the soil temperature at 10 cm depth increased from 15 to 27^{0}C. the optimum temperature for tuber development was 17^{0}C and it practically deceased as soil temperature increased to 29^{0}C. Both shoot growth and yield of rice decreased when temperature was lower or higher than 32^0/20^{0}C (Chaudhary and Ghildayal, 1971).

Root growth: Most of the crops the optimum temperature for growth varies between 20 – 30^{0}C. Lal (1974) found that maize seedling roots ceased to grow beyond 36^{0}C, while the optimum is 26^0 C (Mosher and Miller,1972).

Nutrient availability and uptake: Soil temperature influences soil nutrient aspect in many ways, its availability, absorption and assimilation in plant parts for further metabolic activities. It also influences soil microbial activity and there by decomposition rate. Microbial activity is restricted below 10°C, the optimum being between 18 to 30°C (Zimenko and Revinskya (1972). Thiaglingam and Kanehira (1973) observed 25°C as the optimum soil temperature for nitrification and for the growth of nitrogen fixing bacteria. Knoll *et al* (1964) found increased corn growth and N, P, Fe and Mn uptake with increasing soil temperature. In rice uptake of these nutrients was maximum at 32/20°C.

Global Warming- Are we serious about soil warming?

The above over view on the impact of soil temperature clearly establishes its importance. The impact of climate change vis-a-vis global warming is now well established and a good account work/ literature is available. However, adequate attention is not paid, particularly under Indian conditions, that how the change in atmospheric temperature influences the soil warming? How does the soil temperature profile will behave? The changes in different soil types, cropping sequences and the management practices are the broad area of interest. Some of the classic old literature provides the glimpses of the effect of soil temperature on various aspect of plant growth. After the era of green revolution, streams of water has flown in the river Ganga. Soil quality with respect to physical, chemical and biological indices has changed over the years that too on deterioration side. Though, interdependent, soil physical health is directly related to change in thermal regime by way change in its conductivity and diffusivity. Soil compaction is one that is more worry some especially under intensively irrigated area with rice-rice as predominant cropping pattern. Long term experiment on effect of irrigation on physico-chemical properties in canal command area has shown a decrease in soil porosity and structural change from granular/crumby to sub-angular blocky. It is the cause of worry as it affects the root penetration. Thus, the subsoil nutrients are untapped, relying more on surface application of fertilizers. This again has led to indiscriminate use. The compaction brings the soil particles closer influencing the thermal regime. With the time researchable issues changed. For sustainability of the productivity and the environment resource conservation technologies are being generated, validated and recommended .Surajt *et al* (2018), studied the impact of tillage and residue management on thermal regime on sandy loam soil and reported that soil temperature was the most regular in no tillage with residue management up to 3 cm depth. The difference in temperature at 10 cm depth was 2.9°C in no tillage+ residue management, 3.8°C in No tillage+ no residue management, 4°C in conventional tillage + residue management and 3.1°C in conventional tillage without residue management. On the other hand Somasundaram *et al* (2018) did not find any significant difference in soil temperature due to tillage on rainfed vertisol. Thus, such findings warrant extensive studies and monitoring of soil thermal regime and establishing its correlation with atmospheric temperature becomes very important.

Correlation Between Atmospheric and Soil Temperature

Soil temperatures are important but rarely considered as indicator of climate change. Zhan *et al* (2019) reported close relationship between air temperature and soil temperature at different depth. The range of increase in air and soil temperature at 0, 20 and 80 cm depth was most obvious in spring, at 320 cm in summer. Mean soil temperature showed decreasing trend with soil depth in autumn and winter. Air temperature was lower than the soil temperature at all depth in autumn and winter. He concluded that such trends can form a basis for assessing the climate change impact on terrestrial ecosystem. Garcia *et al* (2006) reported from their study in Ireland that in the past centaury the rate of soil warming has reached 0.04-0.25^{0}C/ decade. Qian (2011) reported that soil temperature showed a trend of increasing of 0.26 – 0.30^{0}C/ decade in spring at Canada during 1958- 2008. Soil temperatures at 40 to 320 cm soil depth have increased by 9^{0}C during winter, while air temperature has increased by approximately 4- 60 ^{0}C (Zhan *et al*, 2001)). Based on analysis of the long term changes at south China, Chen and Zhou, (Chen and Zhou, 2013) found that the annual variation of air temperature correlated very well soil temperature below 80 cm depth. The rate of increase and the rate of annual mean air temperature were reported as 0.184^{0}C/ decade and 0.8^{0}C, respectively.

Conclusion

The soil thermal regime has a profound influence on plant and soil related activities. The very effect on germination and plant establishment decides the number of plant per unit area. An optimum plant population is the first and vital step towards realization of yield potential of any genotype. Little attention has been made on such aspect. The reports from Odisha are still awaited. The Institute of Agricultural Sciences can take the lead. A well defined research agenda on long term mode with multi disciplinary approach can be a right step towards establishing the institute in an area, not addressed effectively so far.

References

Bierhuiezn, J.E.1973.The effect of temperature on plant growth, development and yield. Proceedings of Uppsala Symposium Plant Response to climatic factors. UNESCO: 88-98

Bengtsson, M., Shen, Y and Oki,T. 2006. A SRES-based gridded global population data set for 1990- 2100. Population and Environment.28: 113-131.

Chaudhary, T.N and Ghildyal,B.P.1970. Temperature in relation to seed germination and seedling development in rice (Oryza sativa. L) IL RISO 19(2): 165-68

Chen, C and Zhou, G.S. 2013. Analysis on variation characteristics of air temperature in Guilin from 1961 to 2010. Acta Ecologica Sinic. 33(7); 2043-2053

Garcia- Suarez, A.M and Butlen, C.J. 2006. Soil temperature at armegh Observatory, Northern Ireland from 1904- 2002. International Journal of Climatology. 26(8): 1075-1089

IPCC(International Panel on Climate Change). 2007. Climate Change. 2007: The scientific basis. Contribution of Working Group I to the fourth Annual Assessment report of the IPCC, Cambridge University Press, Cambridge, UK

Mosher, P.N and Miller, M.H.1972. Influence of soil temperature on the geotropic response of corn roots. Agronomy Journal, 64: 459-462

Rao, G.D and Sinha, S.K.1994. Impact of climate change on simulated wheat production in India. In: Implication of climate change for international agriculture: crop modeling study.

Rosegrant, M.W and Cline, S.A.2003. Global food security: Challenges and policies. Science 302, 1917-18

Sinha, S.K and Swaminnathan, M.S.1991. Deforestation, climate change and sustainable nutrition security: A case study of India. Climate change

Somasundaram, J., Sinha, S.K., Mohanty, M., Chaudhary, R.S., Hati, K.M., Singh,R.K., Biswas,A.K., Shukla, A.K., Dalal, Ram and Patra, A.K. 2018. Soil hydro-thermal regimes as affected by different tillage and cropping systems in a raifed Vertisol. Indian Society of Soil Science: 66(4): 362-69.

Surajit Mondal., Das, A., Pradhan,S.,Tomer, R.K., Behera,U.K.,Sharma, A.R., Paul, A and Chakraborty,D. 2018. Impact of tillage and residue management on water and thermal regimes of a sandy loam soil under pigeonpea-wheat cropping system. Indian Society of Soil Science. 66(1): 40-52

Thaiaglingam,K. and Kanehira,V. 1973. Effect of temperature on nitrogen transformation in Havanion soils. Plant and Soil, 38: 117-89

Zhan, M.J., Xia, L., Zhan, L and Wang, Y.2019. Recognition of changes in air and soil temperature at a station typical of china's subtropical monsoon region (1961- 2018). Advances in Meteorology, 20; 19

Zhang, T., Barry, R.C., Gilichinsky,D. Bkyhovets, S.S Sorokovikov, V.A and Ye, J. 2001. An amplified signal of climatic change in soil temperature during the last decade at Irkutsk, Russia. Climatic Change, 49(1-2): 41-76.

Zhang, Y and Chen, W.Y. 2005. Soil temperature in Canada during twentieth century: Complex response to atmospheric climate change. Journal of Geographical Research 110 (3): 3112.

Zimenko,T.C and Revinskya,L.S.1972. Effect of moisture content and temperature on the Activity of micro-organisms in peatbog soils. Microbiologia, 41: 891-95

37

Soil Health
Emerging Aspects, Issues and Suggestions for Farmers Welfare

S.L. Babu and T. Srichandana

Introduction

Soil ecosystems are the foundation of human life support systems. Soil erosion, loss of soil organic matter and nutrient depletion are among the leading contributors to impaired soil health, reduced crop yields and poverty. Realizing the significant role that soils play in food and environmental security and in abating climate change, the United Nations declared 2015 as the Year of Soils. The International Union of Soil Sciences (IUSS), in 2002, proposed to observe the 5th December as "World Soil Day" [WSD] to create awareness among all stakeholders on the importance of soil as a critical component of the natural system and as a vital contributor to human commonwealth through its contribution to food, water and energy security and as a mitigation of biodiversity loss and climate change. These occasions remind all stakeholders associated with agriculture all over the world to put in all efforts to promote scientific management of soil resources for soil protection, conservation and sustainable productivity. The Governments to achieve this goal can [i] seek global technical cooperation and encourage investment in R& D [ii] promote targeted soil research and development focusing on identified gaps and priorities and synergies with related productive, environmental and social development actions [iii] enhance the quantity and quality of oil data and information: data collection (generation), analysis, validation, reporting, monitoring and integration with other disciplines [iv] harmonize methods, measurements and indicators for the sustainable management and protection of soil resources. Last but not the least, to initiate policy and programs to educate farmers and create awareness among them to promote regenerative landscape and integrated management of soil, water, vegetation and biodiversity to enable sustainable agricultural production that is good for the environment and farm profits.

Soil Health Issues and Management

Indiscriminate fertilizer use is driven by the compulsion to maximize short-term yields. The Green Revolution, probably the greatest achievement of post-independence India, heralded an era of food sufficiency riding on the use of chemical fertilizers. Now, 50 years on, soil health is rapidly declining. There is ample evidence to show that indiscriminate use of fertilizers the major cause of deteriorating soil health. Indian farmers apply around 66 million tonnes of fertilizers every year, which accounts for a significant share of India's imports and subsidies. This has resulted in soil degradation and loss of fertility. Soil degradation is estimated to be severely impacting 147 million hectares of cultivable land in India, causing a successive deterioration in its productive capacity. In recent years, experts have witnessed a worrying sign of declining total factor productivity and compound growth rates of major crops.

In recent years, experts have witnessed a worrying sign of declining total factor productivity and compound growth rates of major crops. In several agricultural regions, there has been observed a gap between nutrient demand and supply including decline in organic matter status, deficiencies of micro-nutrients in soil, soil acidity, salinization and sodification. If we do not take this disturbing trend into account and start acting, India might be saddled with vast swathes of land rendered infertile by lack of sagaciousness and long-term thinking. Experts say one of the ways forward is to make agriculture more sustainable and reviving age-old practices of soil regeneration, while balancing it with judicious use of agrochemicals. The agrochemical industry must invest in producing organic biological products that help improve soil health. The major issues related to declining soil health are discussed in subsequent manner.

Soil Erosion

Globally, about 24 billion tonnes of fertile soil is lost annually through water erosion (FAO, 2011). A recent national database on land degradation in India shows that 120.7 Million hectare (M ha) or 36.7% of the total arable and non-arable land surface of the country suffers from various forms of degradation with water erosion being its chief contributor in 83 M ha (68.4%) (NAAS, 2010). Water erosion is the major threat to soil health/quality and runoff water quality. It results in loss of organic carbon, nutrient imbalance, soil compaction, decline in soil biodiversity, and contamination with heavy metals and pesticides. Annual soil loss rate in our country is about 15.35 tonnes per hectare, resulting in loss of 5.37 to 8.4 Million tonnes (Mt) of nutrients, reduction in crop productivity, occurrence of floods/droughts, reduction in reservoirs capacity (1 to 2% annually), and loss of biodiversity (Sharda and Ojasvi, 2016).

i. **Salinization and Alkalization:** The net irrigated area in India has increased from about 22 M ha in 1950 to about more than 51 M ha at present. In most of the expansion, the area is increased under canal irrigation that leads to rise in groundwater Table resulting in the soil deterioration through accumulation of salts. However, with reclamation efforts, the area of 7 M ha soils suffering from salinization or alkalinization has decreased to 6 M ha. The alkali soils in general are characterized by high soil pH (up to 10.8), high exchangeable sodium per cent (ESP) up to 90, low organic carbon, poor infiltration and poor fertility status. These soils are dominated by sodium carbonate and sodium bicarbonate salts. On the other hand, the saline soils have higher electrical conductivity (> 4dS m-1), low ESP (< 15) and low pH (< 8.5). The dominant salts in saline soils include chlorides and sulphates of Na, Ca and Mg. Similarly, 25% of the groundwater resources in the country are saline and brackish. Certain states like, Rajasthan and Haryana are endowed with 84 and 62% of poor quality groundwater, respectively. Continuous use of such waters in providing irrigation to agricultural crops is bound to increase the problem of salinity and sodicity in India.

ii. **Acidity:** The largest areas covered by acid soils in India belong to laterites and various latosolic soils e. g. Ferruginous red soils, ferruginous gravelly red soils, mixed red and black, or red and yellow soils. It is reported that about 6.98 M ha area is affected by acid soils; which is about 9.4% of TGA (Maji, *et al*, 2008). These acids soils are formed due to excessive leaching of cations with high rainfall, resulting in lowering of pH and loss of soil fertility. Based on the range of pH values, the acidic soils have been classified as strongly acidic (pH < 4.5); moderately acidic (pH4.5–5.5); slightly acidic (pH5.5–6.5) and non-acidic(pH>6.5). The management of acid soils includes (a) addition of lime and/or other chemical amendments to correct the acidity and manipulate the agricultural practices so as to obtain optimum crop yields, (b) grow acid tolerant crops and cultivars/varieties and to supplement nutrients through suitable carriers, and (c) water management and other agronomic practices. Lime requirement, estimated on the basis of exchange acidity and percentage bases at uration of soil, approximately ranges from 3.5 to 15tha^{-1}. Acid soils pose typical water management problems, which are mostly associated with physical and chemical properties of soils. Kaolinite dominated light textured acid soils have very high saturated hydraulic conductivity leading to heavy percolation losses. This can be 2 controlled by compaction. Light and frequent irrigation practice helps in enhancing water and nutrient-use efficiency on these soils.

Poor aggregate stability is also one of the major constraints responsible for poor water-use efficiency. Stability of aggregates can be achieved by application of compost, paddy straw, and green manuring. Liming followed by light irrigation is the most effective technique, which helps improve the chemical and biological properties of acid soils and increase water-use efficiency.

iii. **Physical degradation:** The physical deterioration of soil health through water logging, submergence, flooding, soil compaction, crusting, poor infiltrability and impedance to root penetration become limiting factors for crop production. The improper and untimely tillage, lack of return of crop residues to soil or inadequate availability of energy for deep tillage or for conservation agricultural practices are some of the factors responsible for these problems. They all contribute towards loss in grain productivity. Water logging alone results in annual loss of 1.2 to 6.0 Mt of grain in India. (Bradon and Kishore, 1995).

iv. **Diversion of agricultural lands to other purposes:** About 7% of the geographical area, for which land-use statistics is available, is used for nonagricultural purposes, this area is estimated to be increasing at the rate of 0.3 million ha/year. About 20,000 ha of fertile land are being converted into wasteland of poor soil health every year in different parts of the country for brick making for various kinds of construction work. This amounts to the loss of 540 million t/year of fertile soil. India is the second largest brick producer in the world after China. Traditional brick making not only involves removal of top soil but also contributes to GHGs emission. It is estimated that currently 0.7 M ha of agricultural land is being utilized by this industry alone (Nath *et al.,* 2018). The complete process contributes to 40.65 to 42.64 Tg CO2 every year. Use of this sub soil after the topsoil removal can reduce crop yields by 60-90%. Since the Govt. of India is laying emphasis in "Housing for all" and India would have more than 50% urban population by 2050, this situation is likely to worsen and needs to be addressed on priority.

v. **Chemical degradation:** Chemical degradation of soil health is due to i) salinization/alkalinization ii) acidification iii) soil toxification through chemicals, and iv) depletion of nutrients and organic matter and other nutrient input related issues. Around 6.74 M ha are under salt affected soils comprising of 3.79 M ha under high sodicity ($pH > 9.5$) and about 3 M ha (including 1.25 M ha coastal salinity) are under salinity (NAAS,2010a). Soil toxification through chemicals is increasing with greater urbanization. More and more municipal and industrial wastes are being dumped into the soil with heavy metals having carcinogenic effects. A study at the Indian Institute of Soil Science, Bhopal indicated high concentration of heavy metals (Cd, Cr, Cu, Pb, Ni and Zn) in composts manufactured in many cities of India from mixed municipal solid wastes. These heavy metals may accumulate in soil with repeated applications. Extractive farm practices such as higher use of chemicals and burning of crop residues are degrading Indian soil and jeopardizing the health of its citizens, says Rattan Lal, soil scientist, and winner of the 2020 World Food Prize. According to Lal, who is currently the director of Carbon Management and Sequestration Center at The Ohio State University, India needs to formulate a national soil protection policy and pay farmers for ecosystem services (Sayantan Bera, 2020). Soil organic matter content in most cropland soils of north western India and elsewhere is often less than 0.5%. This leads to low and stagnating crop yields.

vi. **Poor soil fertility due to imbalanced nutrient use:** Soils of about 59, 36 and 5% area are low, medium, high in available N, respectively. Similarly, soils of about 49, 45 and 6 per cent area are low, medium and high in available P, respectively; and soils of around 9, 39 and 52% area are low, medium and high in available K, respectively (Chaudhari *et al.*, 2015). Not only the inherent soil fertility is poor and the nutrient input is low but also there is growing evidence of increasing deficiency of P and K, aggravated by the disproportionate/imbalance application of higher doses ofN in relation to P and K (Tewatia *et al.,* 2017). The micronutrient deficiency in crops is growing rapidly both in extent and intensity and as per assessment made under All India Coordinated Research Project on 'Micro and Secondary Nutrients and Pollutant Elements in Soils and Plants' nearly 49, 15, 6, 8, 11 and 33% samples were found to be deficient in zinc, iron, manganese, copper, molybdenum and boron, respectively, across the country and hence all contribute towards poor soil health (Shukla *et al.*, 2012). The nutrient imbalance in soils could have much impact on crop production and environmental protection. Both of the N imbalance and the N losses can be improved greatly, without sacrificing the crop yield, such as balanced and integrated use of fertilizers and organic manures, which are effective in increasing crop yields, nutrient use efficiency and minimizing environmental impacts.

On-going Government Policies for Sustainable Soil and Nutrient Management

a) Entire urea is now being sold as neem oil coated urea with an objective to enhance the N use efficiency and to prevent diversion to non-agriculture usage. The Governmentt's policy of making only neem-coated urea to be marketed in India has certainly led to a decrease in consumption by 10% and enhanced yield of various crops by 6-17%. The actual benefits need to be quantified both in terms of actual reduction in use, yield enhancement, leaching nitrate, and reduction in GHGs. Urea is the only fertilizer that remains under statutory price control and its maximum retail price is fixed by the government. The difference between production cost and the administered concessional price is allowed as subsidy.

b) The Government of India through National Mission on Sustainable Agriculture (NMSA) and 'National Action Plan for Climate Change (NAPCC) for India' is promoting sustainable use of natural resources. Besides, the National Water Mission and National Mission for a Green India under NAPCC have direct relevance to management of natural resources and rainfed areas. The Government under NMSA has also included a component of Reclamation of Problem Soils viz., saline, alkali and acid soils.

c) Under the component of Soil Health Management (SHM) of NMSA the government is promoting soil test-based balanced and integrated nutrient management through setting up/strengthening of soil testing laboratories, establishment of bio-fertilizer and compost units, use of micronutrients, trainings and demonstrations on balanced use of fertilizers.

d) A National Mission on Soil Health Card has been launched under which a Soil Health Card is being issued to each farmer by the soil testing laboratory located in each district of the state. This has been done with an objective to provide soil test-based fertilizer recommendations to all the farmers in the country. Soil Health Card provides two sets of fertilizer recommendations for six crops including recommendations of organic manures. Farmers can also get recommendations for additional crops on demand. They can also print the card as their own from SHC portal. SHC portal has farmers database of both the cycles and is available in 21 languages for the benefit of the farmers. So far 6954 villages have been identified by the States/UTs in which against the targetof 26.83 lakh samples, 20.18 lakh samples have been collected, 14.65 lakh samples have been analyzed and 13.54 lakh card4 have been distributed to farmers. Apart from this 246968 demonstrations and 6951 farmer melas have been approved for States/UTs. (https://pib.gov.in 2020).

e) Paramparagat Krishi Vikas Yojna (PKVY) was launched by the Govt. of India in 2015 as a component of Soil Health Management (SHM) as an initiative to promote organic farming in the country. Cluster approach is being followed to implement this scheme with every cluster consisting of 50 farmers each having one acre for conversion to organic farming. The target was to have 10,000 such clusters in the country over a three-year period.

Suggestions to Overcome Soil health Issues for Farmer's Welfare

a) Implementation of technologies to arrest water and wind erosion developed by the ICAR Institutes and State Agricultural Universities may be reviewed and strengthened.

b) In order to maintain the quality and authenticity of soil test values, there has to be a strong Monitoring mechanism for cross- checking or referencing of soil testing data so as to make the programme viable. For this, some reference laboratories need to be identified and strengthened.

c) Biofertilizers (BF) along with organics has to be an integral part of nutrient management. Deployment of new microorganism or consortia of microorganisms and quality control of Biofertilizer packets supplied at farmer's doorstep should be of paramount importance

d) The deficiency of secondary and micronutrients can be overcome to a great extent by fortification of the presently manufactured N/P/NP/NPK fertilizers ensuring supply of micro and secondary nutrients.

e) Introducing and providing carbon-credit or incentives to the farmers practicing conservation agriculture for carbon sequestration and greenhouse gas mitigation may be a good option. Introducing and providing carbon-credit or incentives to the farmers practicing conservation agriculture for carbon sequestration and greenhouse gas mitigation may be a good option.

Covid-19 has aggravated the problem of food and nutritional insecurity in India and elsewhere by disruption of the food production and supply chain. Contamination of water and pollution of air aggravates human health problems. We must strengthen local food production systems and enhance their resilience. Agriculture must be nutrition-sensitive. Food must be nutritious, enriched in protein and other micro nutrients. Our ancient Vedic culture states that "When food is right, medicine is of no need; when food is wrong, medicine is of no use" and right food comes from crops grown on a healthy soil in a clean environment.

References

Chaudhari, S.K., Biswas, P.P., Abrol, I.P. and Acharya, C.L. (2015) Soil and nutrient management policies. In: State of Indian Agriculture–Soil (Eds.)

H. Pathak, S.K. Sanyal and P.N. Takkar. National Academy of Agricultural Sciences, New Delhi, India, pp. 332-342 Environmental degradation in India. In: Fifty Years of Natural Resource Management Research (Eds.) G.B. Singh and B.R. Sharma, Indian Council of Agricultural Research, New Delhi.

Maji, A.K., Obi Reddy, G.P. and Meshram, S., 2008. Acid soil map of India. Annual Report 2008. National Bureau of SoilSurvey and Land Use Planning, Nagpur, India.

NAAS (2010) Degraded and wastelands of India, status and spatial distribution, Indian Council of Agricultural Research, New Delhi and National Academy of Agricultural Sciences, New Delhi. 150 pp.

Nath, A.J., Lal, R. and Das, A.K. (2018) Fired bricks: CO2 Emission and Food Security. Global Challenge. DOI:10.1002/gch2.20170015.

Sharda, V.N. and Ojasvi, P.R. (2016). A revised soil erosion budget for India: role of reservoir sedimentation and land-use protection measures. Earth Surf. Process. Landforms, 41:2007-2023.

Shukla, A.K. and Behera, S.K., Shivay, Y.S., Singh, Pooja and Singh, A.K. (2012). Micronutrient and field crop production in India: A review, Indian Journal of Agronomy, 57 (3rd IAC Special Issue):123-130.

Tewatia, R.K., Rattan, R.K., Bhende, Shital and Kumar, Lalit (2017). Nutrient use and balance in India with special reference to phosphorus and potassium. Indian Journal of Fertilizer, 13(4):20-31. www.pib.nic.in

38

Impact of Climate Change on Horticultural Crops and Its Management

P.N. Mohapatra and D. Sahoo

Introduction

Seasonal or annual fluctuation in weather patterns is termed as climate variability. However, climate change is a change in the mean state of weather statistic over three decades of time period. Climate change is exacerbating the challenges faced by the agriculture as well as horticulture sector. For last 10 years, Asia is facing the serious threats from climate change and variability such as tsunami, floods and droughts. Extreme events are expected to have large negative impacts on crop productivity. The rise in temperature and reduction in rainfall under semi-arid environments are likely to reduce production of fruits, vegetables and flowers in the coming 20 years. The production and quality of fresh fruit and vegetable crops can be directly and indirectly affected by exposure to high temperatures and elevated levels of carbon dioxide and ozone. The rise in atmospheric CO_2 levels due to global climate change and horticultural practices has both direct and indirect effects on secondary metabolite synthesis in plants. Temperature increase affects photosynthesis directly, causing alterations in sugars, organic acids, flavonoid contents, firmness, and antioxidant activity. Rise in atmospheric carbon dioxide levels persistently affected post-harvest quality causing sugar content reduction in potatoes and tuber malformation incidence of common scab. Elevated atmospheric ozone can result in the decline of the photosynthetic rate, growth, biomass accumulation, increment in vitamin C content, and potential reduction in volatile ester emissions in strawberries. Tomatoes exposed to ozone concentrations ranging from 0.005 to 1.0 μ mol/mol have resulted in a transient increment in β-carotene, lutein, and lycopene contents. Regardless of global climate changes, new cultivation and processing techniques are being adopted in order to produce sufficient food supplies to meet the basic nutrient requirements of the growing human population.

Vegetable crops are very sensitive to climatic vagaries and sudden rise in temperature as well as irregular precipitation at any phase of crop growth can affect the normal growth, flowering, pollination, fruit development and subsequently decrease the crop yield. Climate change induces increase in temperature, variation in rainfall pattern and several other extreme weather conditions that adds to pressure on the global agriculture system and leading to global warming. Now global warming is visible in many parts of the world. It is induced by abnormality in climate patterns and accelerated warming as a result catchment-specific hydrologic cycle is affected. Higher temperatures lead to a high rate of evaporation and dry conditions in some areas of the world. Scientists believe that rapid warming in the last several decades is mostly due to human-induced changes in the atmosphere due to deforestation, ground water pollution, soil degradation, population explosion etc. Impacts of climate change are complex as they can be both direct and indirect, the biggest casualty being natural resources such as agriculture, horticulture and its allied fields. Moreover, due to climate change the perennial plants (mostly fruit plants) are at more risk than annuals or seasonals. As a result global climate change has now become a serious threat to the stability of societies, natural and managed ecosystems. Indeed, the mean global surface temperature during last century has been increased by 0.6 °C. Similarly, further increase in temperature of 1.4 to 5.8 °C is being expected by the end of current century. Impact of climate change on horticultural crops is mainly based on increase in carbon dioxide concentration, temperature, drought, salinity, flooding etc.

Impact of Climate Change on Fruit Crops

Influence of Elevated CO_2

- Increase in atmospheric CO_2 is reported not only to improve the yield but also to alter the quality of the produce.
- The quality (carotene, starch and glucose content) starchy fruits increased in elevated CO_2 conditions. Under high CO_2 level, the water use efficiency of crops is reported to increase.
- In coconut, areacanut and cocoa are reported to benefit due to CO_2 concentrations. Studies indicated that the elevated CO_2 (550 and 700 ppm) increased growth of coconut seedling by 20 to 30%.

Effects of High Temperature

Each fruit crop requires a specific temperature range from minimum to maximum within which a crop will survive. Cold waves during December 2002 – January 2003, caused considerable damage to fruit crops like mango, guava and papaya in India. Even though elevated CO2 is expected to cause positive impacts, however, this may be nullified by increased temperature and less water availability and thereby can decrease production under the current level of management. Flowering and fruit set are adversely affected by high temperature as in mango and mandarin. Citrus fruits suffer a production loss when temperature rise over 37°C are experienced.

a. Growth and Development

In summer the leaves, new shoots, branches, flowers and fruits are exposed to sun and high temperature. High temperature is generally more harmful when accompanied by low humidity. This causes excessive transpiration and wilting of leaves and twigs. The ability of plants to withstand such condition varies greatly with species; e.g. banana is more sensitive than guava plants. Young trees are affected more than the old plants. Sun burn of leaves, fruits (pineapple) and bark is sometimes a serious factor. Plants planted in a east and south aspect are more affected than in other directions as they are exposed to sunlight for longer time.

b. Flowering and Pollination

Weather conditions during flowering, pollination and subsequent fruit growth determine the production quantity and quality.

A cool and dry period, generally in winter, that slows or stops vegetative growth is essential in inducing flowering. On the other hand, low temperature (4to11°C), high humidity (> 80 %) and cloudy weather in January delayed panicle emergence, low temperatures during inflorescence development reduce number of perfect flowers. In citrus, ultimately winter rains promoted vegetative flushes instead of flowering flushes while, dry spell during flower emergence and fruit set affected flower initiation and aggravated pest (Psylla) incidence.

Wash out of pollen grain during flowering has been reported by various workers from different parts of the world. During rainy days pollinating insects remain dull and effective pollination cannot occur and affects fruit setting. After continuous moist weather severe attack of mango hopper certain fungi cause heavy shedding of flowers and fruits and often results in crop failure

Pre-harvest low light intensity due to cloudy weather reduced content of ascorbic acid and sugar in fruits. Heavy rains during harvest period adversely affected production and quality of cashew nuts due to nut germination. Extension of monsoon rains beyond October in west coast delays bud break and flowering.

c. Yield

Apple productivity declined up to elevation of 1500m above MSL to tune of 40 – 50 % due to warmer climate resulting in lack of chilling requirement during winter and warmer summers in lower elevations resulting into shifting of apple production to higher elevation (2700m above MSL). High temperature and moisture stress resulted in sun burn and cracking in apples, apricot, cherries and litchi, dehydration injury to panicles and low fruit set in mango.

d. Pests and Diseases

Under the changing climatic situations, existing fungal pathogen, bacteria, viruses may cause more damage. Some of the minor pests may become major pests in future. Added to these, vector population may increase and new pathogens may emerge. Increase in temperature with low RH (< 60 %) reduced leaf spot diseases in banana in the sub humid region. Bright sunshine hours less than two hours per day triggered incidence of inflorescence blight in cashew etc. Tea mosquito bug populations and damage score in cashew were negatively correlated with minimum temperature (15 – 20°C), RH (85 – 90 %) and positively with bright sunshine hours (9 -10hours).

Effects of Low Temperature

In the hills and plains of north India frost occurs either during winter or in early spring. Low temperature effects due to freeze and frost occurrence in these area. Low temperatures (Freezing) occurring either during the rest period of fruit plants or after wards, cause damage to roots, stems and buds. At lower temperatures, water is withdrawn from protoplasm and ice is formed in the intercellular spaces. If this process proceeds beyond critical limits the protoplasm of the cells disintegrates and dies from loss of water.

Frost causes damage to the newly opened blossoms of fruit trees, to young growth and newly set plants. Frost occurs when there is sufficient moisture in the air and the temperature of the exposed surfaces fall below 32°F. Important types of injuries due to occurrence of frost observed in fruit plants are (1) Black heart—Inner wood of nursery fruit plants becomes dark; (2) Splitting of bark—this may be extended to trunk and branches; (3) Killing of shoots and young branches— if there is early growth in spring and flowering in fruits like mango, flowers may be killed while in others like citrus and papaya even fruits are injured; (4) Frost damage— evergreen tree are damaged heavily but deciduous fruit trees shed their leaves and not normally damaged. In young fruit plants, the damage is very severe.

Impact of Climate Change on Vegetable Crops

Climate change is the primary cause of low production of most of the vegetables worldwide and reducing average yields for most of the major vegetables. Moreover, increasing temperatures, reduced irrigation-water availability, flooding, and salinity will be the major limiting factors in sustaining and increasing vegetable productivity. Under changing climatic situations crop failures, shortage of yields, reduction in quality and increasing pest and disease problems are common and they render the vegetable production unprofitable. The issue of climate change and climate variability has thrown up greater uncertainties and risks, further imposing constraints on vegetable production systems.

The changing patterns of climatic parameters like rise in atmospheric temperature, changes in precipitation patterns, excess UV radiation and higher incidence of extreme weather events like droughts and floods are emerging major threats for vegetable production in the tropical zone. Vegetable crops are very sensitive to climatic vagaries and sudden rise in temperature as well as irregular precipitation at any phase of crop growth can affect the normal growth, flowering, pollination, fruit development and subsequently decrease the crop yield.

Climate change might result in price hike of vegetable crops. Moreover, climate change is fostering the spread of pathogens and the evolution of new strains of insect pests and fungal, bacterial and virus diseases. Challenges ahead are to have sustainability and competitiveness, to achieve the targeted production to meet the growing demands in the environment of declining land, water and threat of climate change, which needs climate-smart vegetable interventions, which are highly location-specific and knowledge intensive for improving production in the challenged environment

Some of the important environmental stresses that affect vegetable crop production are discussed below.

Influence of Temperature

Fluctuations in daily mean maximum and minimum temperature is the primary effect of climate change that adversely affects vegetable production, as many plant physiological, bio-chemical and metabolic activities are temperature dependent. The occurrence of high temperature influences the vegetable production in tropical and arid areas. High temperature causes a significant alteration in morphological, physiological, biochemical and molecular response of the plant and in turn affects the plant growth, development and yield.

- High temperatures in tomato results in bud drop, abnormal flower development, poor pollen production, dehiscence, and viability, ovule abortion and poor viability, reduced carbohydrate availability, and other reproductive abnormalities. Similarly high temperatures above 25°C affect pollination and fruit set in tomato. Moreover, high temperature can cause significant loss in tomato productivity due to reduced fruit set, and smaller and lower quality fruits.
- In pepper, high temperature exposure at the pre-anthesis stage did not affect pistil or stamen viability, but high post-pollination temperatures inhibited fruit set, suggesting that fertilization is sensitive to high temperature stress. High temperature affects red colour development in ripen chilli fruits and also causes flower drop, ovule abortion, poor fruit set and fruit drop in chilli.
- Germination of cucumber and melon seeds is greatly suppressed at 42 and 45°C, respectively besides germination will not occur at 42°C in watermelon, summer squash, winter squash and pumpkin seeds. The temperature fluctuations delay the ripening of fruits and reduce the sweetness in melons. Warm humid climate increase the vegetative growth and result in poor production of female flowers in cucurbitaceous vegetables like ash gourd, bottle gourd, pumpkin that causes low yield. High temperatures would also cause enhanced abscission of flower buds, flowers and young pods and reduce pod production,
- High temperature causes bolting in cole crops, which is not desirable when they are grown for vegetable purpose.

Influence of Drought

Water availability is expected to be highly sensitive to climate change and severe water stress conditions will affect crop productivity, particularly that of vegetables. Drought is a major problem in arid and semi-arid regions, which is the primary cause of crop loss worldwide, reducing average yields for most of the crop plants by more than 50%. Drought stress because of insufficient rainfall or deficient soil moisture might induce various biochemicals, physiological and genetic responses in plants, which severely restricted crop growth.

- The prevalence of drought conditions adversely affects the germination of seeds in vegetable crops like onion and okra and sprouting of tubers in potato.
- The drought condition induces flower abscission in tomato. More than 50% yield reduction was reported in tomato because of water stress

during reproductive stage. It has been suggested that water stress at flowering stage reduces photosynthesis and the amount of photosynthetic assimilates allocated to floral organs and might thereby increase the rate of abscission. Drought stress causes an increase of solute concentration in the soil, leading to an osmotic flow of water out of plant cells. This leads to an increased water loss in plant cells and inhibition of several physiological and biochemical processes such as photosynthesis, respiration etc., thereby reduces productivity of most vegetables.

- Apart from inhibiting the photosynthetic rate through reducing stomatal conductance, drought stress also induces metabolic impairment. The photosynthesis and photosynthetic capacity are reduced during limited water conditions. Further, the biochemical capacity was also affected by the water stress as indicated by a decrease in sucrose phosphate synthase (SPS) and invertase activities, which affect the availability and utilization of sucrose. The SPS is considered to play a major role in the resynthesis of sucrose and sustain the assimilatory carbon flux from source to developing sink. The decreased invertase activity might affect the ability to utilize sucrose and result in reduced ovary growth and reduced concentration of hexoses.

Influence of Salinity

Salinity is a serious problem that reduces growth and productivity of vegetable crops in many salt-affected areas. Excessive soil salinity reduces productivity of many agricultural crops, including most vegetables, which are particularly sensitive throughout the ontogeny of the plant. Physiologically, salinity imposes an initial water deficit that results from the relatively high solute concentrations in the soil, causesion-specific stresses resulting from altered K^+/Na^+ratios, and leads to a buildup in $Na^+$and Cl^-concentrations that are detrimental to plants. Salt stress causes loss of turgor, reduction in growth, wilting, leaf abscission, decreased photosynthesis and respiration, loss of cellular integrity, tissue necrosis and finally death of the plant.

- Onions are susceptible to saline soils, while cucumber, eggplant, pepper, and tomato are moderately sensitive to saline soils.
- Salinity causes a significant reduction in germination percentage, germination rate, and root and shoots length and fresh root and shoot weight in cabbage.
- The combined stress of salinity and heat results in failure of vegetative growth recovery and a consequent reduction in the leaf area index and canopy functioning due to the damage of salt accumulation avoiding mechanism in young expanding leaves of potato.

- Salinity reduces dry matter production, leaf area, relative growth rate and net assimilation rate in chilli. High salt concentration causes a reduction in fresh and dry weight of all cucurbits. These changes are associated with a decrease in relative water content and total chlorophyll content.
- Salt stress causes suppression of growth and photosynthesis activity and changes in stomata conductivity, number and size in bean plants. It reduces transpiration and the cell water potential in salt-effected bean plants. The high salinity levels of soil and irrigation water are known to affect many physiological and metabolic processes, leading to cell growth reduction.

Influence of Flooding

Flooding is another important abiotic stress and cause serious problems for the growth and yield of vegetable crops. The occurrence of flooding conditions normally cause oxygen (O_2) deficiency which arises from a slow diffusion of gases in water and O_2 consumption by microorganisms and plant roots. Most vegetables are highly sensitive to flooding and genetic variation with respect to this character is limited, particularly in tomato. In general, damage to vegetables by flooding is due to the reduction of oxygen in the root zone, which inhibits aerobic processes.

- Flooded tomato plants accumulate endogenous ethylene that causes damage to the plants. The rapid development of epinastic growth of leaves is a characteristic response of tomatoes to waterlogged conditions and the role of ethylene accumulation has been implicated.
- Onion is also sensitive to flooding during bulb development with yield loss up to 30- 40%. These stresses are the primary cause of yield losses worldwide by more than 50% plant and the response of plants to environmental stresses depends on the developmental stage and the length and severity of the stresses.
- Floods can make the spread of water-borne pathogens easier, droughts and heat waves can predispose plants to infection, and storms can enhance wind-borne dispersal of spores.

Pests and Diseases Responses to Climatic Change in Vegetable Crops

Climate change also influences the ecology and biology of insect pests. Increased temperature, in some group of insects with short life cycles such as aphids and diamond back moth, increases fecundity, earlier completion of life cycle. Hence, these can produce more generations per year than their usual rate

- Increased temperature causes migration of insect species towards higher latitudes, while in the tropics higher temperatures might adversely affect specific pest species. High atmospheric temperature increases insect developmental and oviposition rates, insect outbreaks and invasive species introductions.
- Changes in temperature and precipitation regimes due to climate variation may alter the growth stage, development rate and pathogenicity of infectious agents, and the physiology and resistance of the host plant.
- Temperature and frost sensitivity effect the distribution of pathogen species as irrespective of their huge host range the soil borne pathogens such as *Sclerotium rolfsii* and *Macrophomina phaseolina* do not occur in temperate climates due to their high temperature optimum and frost sensitivity.
- The earlier appearance and increase in number of insect vectors of viral diseases due to rise in temperature during winter, results in increasing viral diseases of crops like potato and sugar beet.

Greenhouse Effect and Global Warming

The exchange of incoming and outgoing radiation that warms the Earth is often referred to as the greenhouse effect because a greenhouse works in much the same way. Incoming UV radiation easily passes through the glass walls of a greenhouse and is absorbed by the plants and hard surfaces inside. Weaker IR radiation, however, has difficulty passing through the glass walls and is trapped inside, thus warming the greenhouse. This effect lets tropical plants thrive inside a greenhouse, even during a cold winter but this global warming hampers crop that are grown in open condition.

Thus, greenhouse gas is a gas that absorbs and emits radiant energy within the thermal infrared range. Increasing greenhouse gas emissions cause the green house effect. Basic sources of GHGs emission are electricity and heat production from industry, transportation, agriculture, forestry and other sources.

Adaptation Strategies to Achieve a Sustainable Approach for Horticulture Crop Production to Combat the Impact of Climate Change

Adaptation strategies deal with the negative impacts of climate change and have the objective of reducing the vulnerability of human and natural agro-ecosystems. It aims are to make agro- ecosystem more resilient against natural catastrophes. There are some indigenous principle and practices that have helped shape of climate resilient and sustainable farming all over the world. Following adaptation strategies that have helped to combat horticultural crops from climate resilient and add to sustainable farming all over the world.

Agri-horti-silvi Cultural Systems

Agri-horti-silvi cultural systems involves the deliberate maintenance and planting of trees with field crops and horticultural crops to develop a microclimate that protects crops against extremes. Blending of agriculture with forestry techniques helps to control temperature, sunlight exposure, and susceptibility to wind, hail, and rain. This system provides a diversified range of products such as food, fodder, firewood, timber, and medicine while improving soil quality, reducing erosion, and storing carbon.

Developing Climate-ready Crops

There is necessity to develop and use crop varieties having greater potential for yield and resistance to abiotic stresses in fruits, vegetables and flowers (water logging/flooding, drought, salinity) in order to maintain yield stability. The plants having these desire traits will enable them to perform well under wide range of stresses. The manipulation of gene pyramiding in genetic engineering is necessary to pool up the desired traits in single plant to develop true to type plant. The other breeding approach is transformation of C_3 horticultural plants to C_4.

Crop Diversification

It is replacement of conventional cultivars with newly advanced cultivars in order to get high production in the face changing climate. Crop diversification is known as a defense tool against all abiotic stresses as it exploits efficient utilization of natural resources. Crop diversification also covers the growing of alternate crops in existing intensive cropping systems. This will develop resilience in existing cropping systems against future changing climate.

Adjusting Cropping Season

It is growing of crops in such a manner that they can coincide with prevailing weather conditions of growth season. This will help to crops in completion of their life cycle within preferable environment. Adjustment of planting times has a vital role in minimizing the devastating impact of rising temperature on sterility of reproductive part of crops. It is also necessary to minimize yield instability by avoiding flowering and fruiting from the warmest period.

Resource Smart Technologies

The resource-conserving technologies (RCTs) are those actions which increase resource or efficiencies of inputs and give instant, discernible economic benefits like reduction in cost of production; saving of fuel and labor requirements and timely crop establishment, resulting in higher yields. Resource conservation technology such as no till or zero tillage can permit farmers to grow crop in time prior to onset of hot conditions of pre monsoon season. The resource-conserving technologies have prominent impact on reduction of GHGs emission and adaptation to future changing climate.

Allocation of Crops in Alternative Areas

There is need of hour to recognize the crops and regions which are highly prone to changing climate/variability then relocate those crops in more suitable areas. For instance, increased temperature would have adverse impact on quality of crops having aromatic properties.

Mixed or Inter-cropping

Mixed cropping, also known as intercropping is a system of cropping in which farmers sow more than two crops at the same time. By adopting multi storied cropping in horticulture farmers can maximize land use by cultivating fruits/ plantation crops, vegetables, flowers and spices together and thereby reducing the risk associated with single crop failure. Intercropping creates biodiversity, which attracts a variety of beneficial and predatory insects to minimize pests and can also increase soil organic matter, fumigate the soil, and suppress weed growth.

Development of Environment

Increase the efficiency of natural resources used in Horticulture (water, energy, fertilizer, etc.) to lower the global warming potential of agriculture and to reduce water and air pollution. Stop unsustainable withdrawal of water resources, soil degradation, and soil nutrient depletion. Protect wildlife, biodiversity and other ecosystem services in forest landscapes. Now-a-days, mitigation is more cost effective than adaptation to climate change in developing countries. There are various mitigation options to climate change in agro-ecosystem.

The capturing and storage of atmospheric carbon in agro-ecosystem is known carbon sequestration. Carbon in the soil act as Soil Organic Carbon (SOC) or Soil Inorganic Carbon (SIC), SOC is known as one of the climate change mitigating options. Nevertheless, any small change in SOC may affect atmospheric CO2 concentration. However, management of SOC will play a pivotal role to enhance its amount in soil and also in food security as well as mitigation of future climate. As the higher amount of soil organic carbon will result in higher crop productivity. Resultantly, production of higher biomass would result in more organic carbon inputs into soils and therefore, rise in soil carbon sequestration in the form of SOC. There are some agricultural practices which may lead to enhance soil organic carbon like organic amendments such as green manuring, straw/stubble management and farmyard manure are mainly recommended as an alternate of mineral fertilizers to provide some essential plant nutrients.

Pest Management Strategies

The rising temperature and variations in precipitation have strong correlation with incidence of insects/pests and disease of major crops. It is due to climate change which will have potential effect on pest-host relationship as it affects the pest and host population as well as pest-host interactions. A few potential adaptation packages against pests could be

(i) Development of pest and disease resistance cultivars

(ii) Preferably adoption of integrated pest management (IPM)

(iii) Pest forecasting with advance tools like simulation models

(iv) Pests and diseases management with natural pesticides, bacterial and viral pesticides as well as pheromones to disturb pest reproduction

(v) Bio-agents also have a key role in pest management

Weather Forecasting

Climate influences the distribution of crops over different regions of the world, while weather influences crop production and productivity. Among the various weather elements affecting crops, temperature and rainfall are perhaps the important. It is the reason for the development of special flora and fauna in different climatic zones of the world as seen in the tropical, temperate and frigid region. The influence of weather factors on biotic stress due to incidence of insect pest, disease, weeds, and grazing animals. Weather forecasting will help the farmers to decide the farm operations e.g. pesticides and micronutrient spray, irrigation scheduling, fertilization and harvesting. Farmers g*et al*erts for the management of different abiotic stresses such as heat waves, frost, storms and heavy rainfall.

Precision Agriculture and Nutrient Budgeting with Site Specific Crop Management (SSCM)

This can be done by facilitating more efficient use of fertilizers, can reduce emissions associated with excess application as well as reduce cost. Follow practices that improve N-use efficiency by adjusting application rates based on precise estimation of crop needs (e.g., precision farming); using slow-release fertilizer forms or nitrification inhibitors; avoiding time delays between N application and plant N uptake (improved timing); placing the N more precisely into the soil to make it more accessible to crops roots; avoiding excess N applications, or eliminating N applications where possible.

Water Management

Improving soil moisture management is crucial for the sustainable improvement of food production and water supply. Better water management, coupled with improved soil and crop management, can increase crop productivity in rainfed areas with currently low yields by more than double. With climate change and increasing food prices, even more emphasis needs to be placed on addressing water management in dry land horticulture as a key determinant for horticulture crop production and productivity.

Development of Varieties Using Biotechnology

Increasing crop productivity in unfavourable environments will require advanced technologies to complement traditional methods, which are often unable to prevent yield losses due to environmental stresses. With advancement of the science of genetic engineering, it is time now to develop more and more varieties through genetic manipulation having tolerance to environmental stresses. The engineered tomato has a stronger, larger root system that allows the roots to make better use of limited water. The control plants suffered irreversible damage after five days without water as opposed to transgenic tomatoes which began to show water-stress damage only after 13 days but recovered completely as soon as water was supplied.

Improving Stress Tolerance Through Grafting

Grafting vegetables originated in East Asia during the 20th century with the objective of reducing the effects of soil borne diseases like Fusarial wilt, which affects the production of vegetables such as tomato, eggplant, and cucurbits. Nowadays, Grafting is considered as a common practice in vegetable production in Asian countries such as Japan, Korea and some European countries which is an efficient rapid alternative tool to the relatively slow breeding methodology aimed at enhancing environmental-stress tolerance of horticultural crops in general and vegetables in particular. Grafting is one of the promising tools for modifying the root system of the plants as followed for enhancing its tolerance to various abiotic stresses.

- In vegetable crops, grafted plants are now being used to improve resistance against abiotic stresses like low and high temperatures, drought, salinity and flooding if appropriate tolerant rootstocks are used. Because of these beneficial effects of grafting, the cultivation of grafted plants in crops like tomato, eggplant and pepper and cucurbits (melon, cucumber, watermelon and pumpkin) has increased in recent years.
- Melons grafted on to hybrid squash rootstocks were more salt tolerant than the non- grafted melons.
- Grafting temperature-sensitive tomato onto more resistant rootstock cultivars improves plant adaptation to heat-stress conditions. It is reported that grafted plants found to develop better under heat-stress conditions than the ungrafted plants of tomato.

Conclusion

The transition towards climate resilient agriculture practice is a knowledge-intensive learning process involving multiple stakeholders. Realizing climate resilient horticulture practice thus requires a holistic and comprehensive approach of all stakeholders involved, based on good practices such as alignment with and driven by national priorities; incorporating the individual, organizational and enabling environment dimensions and complementing technical with functional competency to strengthen and apply good learning practices Due to the uncertain and dynamic nature of climate change impacts, the module illustrated that this transition towards climate resilient agriculture requires socio-institutional learning processes with a strategic approach to skills development at country level including strong engagement of national and local formal and informal education and training institutions

39

Statistical Assessment on Impact of Climate Change in Different Districts of Odisha

S.K. Mahapatra, S.K. Mohanty and J. Pradhan

Introduction

The state of Odisha highly depends on rainfall for the agricultural production. But rainfall distribution is very erratic and uneven which result heavy floods and droughts in different pockets of the state. Thus, the agricultural production is very unstable. It is present in the eastern region of India which is again located in the subtropical belt of medium pressure. Here hot and dry summer welcomes wet humid monsoon which last about few months. Charming autumn and with short and mild winter entertain the human life. But here climate is characterized by high temperature and medium rainfall. Topography however modifies the local climate greatly. The four meteorological seasons prevailing in the state which are: winter (January to February), pre-monsoon season (March to May), south-west monsoon season (June to September) and north east monsoon season (October-December). The average annual rainfall of Odisha is almost 1482.2 millimeters out of which major amountis being received during June to September. Failure of adequate rainfall causes scarcity of water while excess causes floods in monsoon but normal rainfall controls the yield.

As per United National Framework Convention on Climate Change (UNFCCC) is attributed directly or indirectly to human activity which has bearing on global atmosphere. Essentially the climate change is not same as weather change. Change of weather is seasonal and temporary whereas climate change involves permanency in change of weather. The established meaningof climate is the measure of the average pattern of variation in temperature, humidity, atmospheric pressure, wind precipitation and other meteorological variables in a given region over a long period of time. It is different from weather and describes the short term conditions of these variables in a given region. In narrow sense climate is average weather at least for a period of 30 years as per the World Metrological Organization. The climate as observed is now subject to significant change. The Climate of Odisha as set up since ages and the seasons are monsoon, winter, summer, spring and autumn, occur at constant time and continues for certain period. This climate of the state has been determined due to a number of factors such as location, ocean currents, forests, direction of prevailing winds, shape of land and influenced by human to a great extend. The above factors have changed the climate of the state to a great extent recently. Climate change has become a matter of great concern. Odisha's seasons have all but vanished, its trees have altered their flowering time, and the farmers have changed their farming practices. Not only this, of the six seasons prevalent there seems only two summer and rain that have their effects on Lands of the state. Autumn, spring and winters have slowly vanished from the memory of the people. While summers have become longer, winters have become warmer and rains have shortened from above 120 to 90 days while becoming erratic beyond point.. Climate is one of the main determinants of agricultural production. Throughout the world there is significant concern about the effects of climate change and its variability on agricultural production. Current climate variability and change is predicted to cause increasing global temperature and subsequently impact the rainfall. Agriculture is one of the most climate sensitive sectors as it is continuously and directly affected by temperature and precipitation. Climate changes pose significant economic and environmental risks worldwide.

Climate change affects to food security is in four dimensions, food availability, food accessibility, food utilization and food system stability. It will also have an impact on human health, livelihood assets and food production and distribution channels (FAO, 2008). Due to rising global population size, climate change will challenge agricultural production and food4 security (location of production, supply, volume, quality) and by 2080, agriculture output Least Developed Countries (LDCs) may decline by 20 percent due to climate change and yields could decrease by 15 percent on average, while output in industrial countries is expected to decrease by 6 percent (Masters *et al.*, 2010). It will affect about 200 million people and their families worldwide who live by fishing and aquaculture (Greg *et al.*, 2011). In India, Gross Domestic Product (GDP) may decrease up to 6.2% and agriculture production may decrease up to 24% by 2080 due to climate change (Zhai and Zhuang, 2009; and Zhai*et al.*, 2009). However, variations between storm characteristics (duration and return period) were more pronounced for short- duration, high-intensity events than they were for long duration, low-intensity events. The sizeable variation between historical and future rainfall intensities for short-duration events, however, generally supported the notion that precipitation intensity will be more severe in future decades (Endreny and Imbeah 2009). Gao *et al.* (2002) have applied stochastic hydrology methods to analyze the characteristics of annual inflow evolution of Miyun reservoir. It is obvious from their experiment that the annual inflow decreasing is mainly caused by human activities, such as extreme population increase and land use changes. Modeling the changes of in-flow to reservoir has been studied by Dilini *et al.* (2013) for effective water management. Sahu etal. (2012) made use of an artificial intelligence technique called the self-organizing map (SOM) to perform trend and cluster analysis for the inflows into the flood-control reservoirs of Indiana. Along with SOM, this research also used the Mann-Kendall test and a revised Mann-Kendall test for regional analysis for northern reservoirs of Indiana. Sethy *et al.* (2015) performed a trend analysis for precipitation and inflows time series for Salia river basin of Odisha, India which is draining to Chilika lake using the Mann–Kendall test. In order to assess the impact of climate change ARNO model was used to simulate the inflows into Salia reservoir calibrating the observed inflow Dawson *et al.* (2015) have studied about trends in water quality and quantity for 11 major reservoirs of the Brazos and Colorado river basins in the southern Great Plains. Brown (2008) has foundout that according to United Nation Population Fund's State of the world population 2009 there will be an increase in migration trends which would reach up to 200 million by 2050. Brooks (2003) made an attempt to develop a conceptual framework of risk, vulnerability and adaptive capacity that synthesizes a variety of approaches. By distinguishing between social and biophysical vulnerability the apparent conflict between different formulations of vulnerability in the climate change literature can be resolved. Chandran (2013) has tried to found out that Climate change is a human development issue which undermines expanding human potential, developing capabilities and enlarging freedom. Climate change poses major obstacles to progress in meeting the Millennium Development Goals (MDGs) and maintaining progress raising the Human Development Index (HDI). Climate change is closely linked to the broader sustainable development agenda to reduce poverty, child mortality and morbidity.

Materials and Materials

The study period consists from the year 1970 to 2018. Daily Rainfall data of 40 years for the period 1970 to 2018 were collected from the SRC Odisha (2019), India metrological department (IMD) website (http://www.Indiawaterportal.org/met data). Monthly, Seasonal(S- W Monsoon, Post Monsoon, winter and summer) and annual rainfall data were computed from daily rainfall data. Statistical Parameters such as Mean, standard deviation, Coefficient of variation, Skewness and Kurtosis of these rainfalls were calculated by using SPSS 20.0. The trend was analyzed using non- parametric Mann-Kendall test (Mann 1945; Kendall, 1975). The MK test has been employed by a number of researchers to ascertain the presence of statistically significant trend in hydrological climatic variables such as temperature, precipitation with reference to climate change. The MK test checks the null hypothesis of no trend versus the alternative hypothesis of the existence of increasing or decreasing trend. The regional water resources study was done by analyzing the rainfall data for 30 districts of Odisha for the period of 1970 to 2018.

Variability analysis (Coefficient of variation)- Assessment of rainfall variability through % appears to be quite simple. CV is abbreviated Coefficient of variation which is defined as the Standard deviation divided by the mean value of rainfall. It expresses the variability of rainfall in percentage.

Trend Analysis: A trend is a pivotal change over time exhibited by a random variable. Here, the magnitude of trend in a time series is determined by using regression analysis (parametric test). This method assumes as a linear trend in the time series. In this particular study, the linear regression test was employed. For study purpose we have to work out trend analysis on annual basis as well as seasonal basis.

Student t test: Student t test was pioneered by W.S. Gosset (1908) who wrote under the pen name of Student, and later on developed and extended by Prof. R.A. Fisher.

$$t = \frac{x-\mu}{\frac{s}{\sqrt{n}}} \sim t_{(n-1)} \text{ df}$$

Let $x_1, x_2......x_n$ be the random sample of size 'n' form a normal population with a mean 'μ' and variance 'σ^2' then student's t-test is defined by statistic

Mann Kendal Test: A stochastic process is a statistical process shown in (figure 4) involving a number of random variables depending on a variable parameter. If the variable parameter is the time, then the process becomes a stochastic time series in this analysis the in-flow series. The most important aspect of the stochasticity is stationary.

Table 1: Annual trend and Z values of 30 districts

Sl. No	District	Annual Trend		Test Statistics Value (Z_value)
		5%	10%	
1	Anugul	No	No	0.77
2	Baleswar	Rising	Rising	2.017
3	Bargarh	No	Falling	-1.75
4	Bhadrak	No	No	1.57
5	Bolangir	No	Falling	-1.85
6	Boudh	No	No	0.41
7	Cuttack	Rising	Rising	2.33
8	Deogarh	No	Falling	-1.88
9	Dhenkanal	No	No	1.29
10	Gajapati	Rising	Rising	2.1
11	Ganjam	No	Rising	1.75
12	Jagatsinghpur	Rising	Rising	3.32
13	Jajpur	Rising	Rising	1.97
14	Jharsuguda	Falling	Falling	-2.01
15	Kalahandi	No	No	-0.37
16	Kendrapada	No	No	1.44
17	Keonjhar	No	Rising	0.9
18	Khurdha	Rising	Rising	2.01
19	Koraput	No	No	1.18
20	Malkangiri	Rising	Rising	2.7
21	Mayurbhanj	No	No	1.33
22	Nabarangpur	No	No	-0.86
23	Nayagarh	Rising	Rising	-2.48
24	Nuapada	Falling	Falling	-2.84
25	Phulbani	No	Rising	1.75
26	Puri	Rising	Rising	2.96
27	Rayagada	No	No	0.44
28	Sambalpur	No	No	-1.05
29	Sonepur	No	No	-0.86
30	Sundargarh	Falling	Falling	-2.13

Table 2: Monsoon rainfall trend at 5% and 10% significant level of 30districts

District	June		July		August		September		October	
	5%	**10%**	**5%**	**10%**	**5%**	**10%**	**5%**	**10%**	**5%**	**10%**
Anugul	No	No	No	Rising	No	No	No	No	No	No
Baleswar	No	No	No	Rising	No	No	No	No	No	No
Bargarh	No	No	No	No	Falling	Falling	No	falling	No	No
Bhadrak	No	No	No	No	No	No	No	No	No	No
Bolangir	No	No	No	No	No	No	falling	falling	No	No
Boudh	No	No	Rising	Rising	No	No	No	No	No	No
Cuttack	No	No	Rising	Rising	No	No	No	No	No	No
Deogarh	No	No	No	No	falling	falling	falling	falling	No	No
Dhenkanal	No	No	No	Rising	No	No	No	No	No	No
Gajapati	No	No	No	Rising	No	No	No	No	No	No
Ganjam	No	No	No	No	No	No	No	No	No	No
Jagatsinghpur	No	No	Rising	Rising	Rising	Rising	No	No	No	No
Jajpur	No	No	No	No	No	No	No	No	No	No
Jharsuguda	No	No	No	No	Falling	Falling	Falling	Falling	No	No
Kalahandi	No	No	No	No	No	No	Falling	Falling	No	No
Kendrapada	No	No	No	No	No	No	No	No	No	No
Keonjhar	No	No	No	No	No	No	No	No	No	No
Khurdha	No	No	No	Rising	No	No	No	No	No	No
Koraput	Rising	Rising	No	No	No	No	No	No	Rising	Rising
Malkangiri	Rising	Rising	Rising	Rising	No	Rising	No	Rising	No	No
Mayurbhanj	No	No	No	Rising	No	No	Falling	Falling	No	No
Nabarangpur	No	No	Rising	No	No	No	No	No	No	No
Nayagarh	No	No	No	Rising	No	No	No	No	No	No
Nuapada	No	No	Rising	No	Rising	Rising	Rising	Rising	No	No

District	June		July		August		September		October	
	5%	**10%**	**5%**	**10%**	**5%**	**10%**	**5%**	**10%**	**5%**	**10%**
Phulbani	No	No	Rising	Rising	No	No	No	No	No	No
Puri	No	No	No	No	No	No	No	No	No	No
Rayagada	No	No	No	No	No	No	No	Rising	No	No
Sambalpur	No	No	No	No	No	No	No	NO trend	No	No
Sonepur	No	No	No	No	No	No	No	NO	No	No
Sundargarh	No	No	No		Falling		Falling	Falling	No	

Result and Discussion

In the present study trend analysis of annual, monsoon and temperature has been carried out. For this initially the available rainfall and temperature data was classified in the above period. Further Mann-Kendall test was applied for three significance levels i.e. 1%,5%and 10%.On the basis of Z-statistics of each significance level the trends in of Orissa has been determined. The trends in total rainfall, annual maximum rainfall, monsoon rainfall and total temperature were investigated through the Kendall's test for Orissa. The monthly data were used to compute seasonal and annual time series of total rainfall. The annual and monsoon rainfall and temperature trend at 5% and 10% significant level of 30 districts are shown below.

The economy of Odisha is fully dependent on agriculture. The quantity and spread of rainfall over entire 30 districts are very important in this regard as most of our land is rain fed. It has been assessed that so many hectares cultivable land can be brought under irrigation through major and medium irrigation projects. By end of March 2020, irrigation facilities of 22.32 lakh hectares have been created. Still a large area remains un- irrigated and agriculture in this area totally depends on the rain. The information on quantification and its trend will definitely help planning and scheduling of our agriculture.

Our state has seven majors and 45 medium irrigation storage structures comprising a live storage capacity of 153246 ham. Further mega lift irrigation storage is being initiated to improve the irrigation scenario. This is something interesting to note for planning soil and water conservation measure including water harvesting and also for planning irrigation to summer crops. The regression slope of annual and monsoon rainfall is almost close and have same pattern. Test of significance (t-test) at 5 % significance level was carried out on the regression slopes for the cases studied. Study revealed that month of August is very critical for kharif crop point of view in this region and rainfall of this particular month is decreasing significantly, hence there is need to harvest more rainfall of July month through construction of different in-situ and ex-situ water conservation and storage structures. This stored water can be useful for providing lifesaving irrigations during long dry spells because the numbers of rainy days were also found to be reduced significantly during monsoon season. Moreover, farmers need to make some changes in their cropping pattern to cope with changing rainfall pattern like switching over to less water demanding crops, such as pulse, millets etc. The trend at monsoon temperature showed that most of districts have no trend but Rayagada, Phulabani, Koraput, Malkangiri, Nayagarh and Puri are shown rising trend at 5% and 10% significant level. The reservoirs show a delayed and decreased inflow during the month of June, which compels for starting of cropping season, may be shifted to late June or early July. Creation of irrigation potentials is highly essential in order to face the varying rainfall scenario.

References

Aggarwal PK. (2007) Climate change: implications for Indian agriculture. Hydrology Review. Roorkee: Indian National Committee on Hydrology. 22, 37–46

Amienn I.P. Rejekiningrum and Pramudia *et al.* (1996), Effects of interannual climate variability and climate change on rice yields in Java, Indonasia‘, Water, Air, and Soil Pollution 92,pp.29-39.

Annual report on natural calamities (2011-12) Government of Odisha

Arora M., Goel NK. & Singh P. (2005) Evaluation of temperature trends over India. Hydrol. Sci. J. 50(1), pp. 81-93.

Bates B. C., Kundzewicz ZW., Wu S., and Palutikof JP. (eds) (2008) Climate Change and Water. Technical Paper of the Intergovernmental Panel on Climate Change.IPCC. Secretariat, Geneva: 210pp..

Brekke LD, Norman LM, Kathy EB, Quinn NW, Dracup JA (2004) Climate change impacts uncertainty for water resources in the San Joaquin River basin, California. J Am Water Res Assoc 40: 149-164.

Brown RA., and Rosenbergn NJ. (1999). Climate change impacts on the potential productivity of corn and winter wheat in their primary United States growing regions." Clim. Change, 411, pp.73-107.

Brumbelow KA; and Georgakakos A.(2007) Consideration of Climate Variability and Change in Agricultural Water Resources Planning , Journal of Water Resources Planning and Management, . 133, (3).

Burn DH (1994) Hydrologic effects of climatic change in west-central , Canada. Journal of Hydrology, 160 (1) .53-70.

Cain, N.L, Haasz, D., Henges-Jack, C., Kiparsky,M., Hunt, C., Moench, M., Palaniappan, M., Srinivasan, V. and Wolff, G.H (Eds.),(2004) The World's Water 2004-2005, Washington DC, Island Press, pp.177.

Christensen JH., Hewitso B., Businoc A. and Chin, A. (2007) Regional climate projections. In: Solomon, S. (ed.) Climate Change 2007: The Physical Science Basis. Contribution of Working Group 1 of the Fourth assessment Report of the Intergovernmental Panel on Climate Change. Cambridge University Press, Cambridge: 840-84.

Climate Change and Water. Technical Paper of the Intergovernmental Panel on Climate Change, IPCC Secretariat, Geneva, pp 210. Pak Sum Low (2005) Climate Change and Africa. New York; Cambridge University Press,pp.30.

Danodia, A and Sastri, (2015) ASRAS Impact of rainfall variability on water resources of Chhattisgarh state with special reference of mannkendall test statistics on the context of climate change, Eco. Env. & Cons. 21 (4) :125-132.

Dash SK., Jenamani RK. & Shekhar M. S. (2004) On the decreasing frequency of monsoon depressions over the Indian region. Current Sci. 86, 1404-1411.

Gosain, AK., Rao, S. & Basuray, D. (2006). Climate change impact assessment on hydrology of Indian river basins. Current Sci. 90,346-353

Lal, M. (2001) Climatic change implications for India's water resources. J. Ind. Water Resour. Soc. 21, 101–119.

40

Combating Drought with Antitranspirants

B.K. Dadrwal, A. Babu, K. Doodhwal, P. Choudhary and U.P. Singh

Introduction

The impacts on the various facets of life – agriculture and food processing – need no longer be addressed in a way that is well established. Nearly every aspect of the planet has shown its effect at different rates, both positively and negatively. As a subtropical region, India is also witnessing different variations in weather and an increase in temperature that affect farming. Only in recent years has the potential to reduce transpiration from plants by the use of chemicals been further explored. These antitranspirants can either control perspiration by filming the leaf surface or by manipulating the aperture of the stomach. Antitranspirants are the materials or chemicals that reduce the water loss from plant leaves by reducing stomata size and number. During transpiration, almost 99 per cent of the water consumed by the plant is lost.

Antitranspirants are chemical compounds, their role is to harden plants to stress, as a method of reducing the salinity-related effect of drought. In this respect, reducing transpiration can play a useful role in preventing the luxurious loss of water to the atmosphere through stomata. Recently, efforts have been made to find substances which reduce transpiration, i.e. antitranspirant materials, when applied to the plant. However, since the transpiration and photosynthesis processes involve the passage of water vapor and carbon dioxide through the stomata, both processes may be affected when the application of antitranspirants narrows or curtails those stomata. Antitranspirants are off:

Film-forming Type

Antitranspirant film forming on the surface of the leaf forms a colorless film which reduces the transpiration rate but has no effect on the exchange of gasses. Film analysis on the surface of the leaf resulted in the formation of film on the surface of the leaves reducing transpiration to a greater extent having a slight effect on growth. Past studies have shown that the antitranspirant forming film is more effective in increasing grain yield and that photosynthesis in both adverse and favorable conditions. The higher leaf turgor in antitranspirant -sprayed plants is consistent with the notion that AT film decreases water loss and allows sustained turgor maintenance under WD (water deficit) conditions In water stress situation leaf maintains its turgidity by applying antitranspirant on it and minimizing water loss in stress. e.g. Plastic films, waxes, silicone oils

Stomatal Regulating Type

Most of the antitranspirant acts like closer stomatal compound when applied over leave. Some low concentration fungicides such as phenyl mercuric acetate (PMA) and herbicide such as Atrazine serve as an anti transpirant by inducing stomatal closure. These could reduce PMA photosynthesis to decrease transpiration compared with photosynthesis. Work has shown that certain fungicides, including phenyl mercury acetate, mercurized copper oxychloride, and copper oxychloride, have decreased the transpiration of tomato seedlings and potato plants. The highly effective compounds tended to be many effective stomatal controlling compounds, and even their concentrations used, the percentage decrease in transpiration and stomatal condition; phenyl mercuric acetate, 8- hydroxyquinoline sulphate, and decenylsuccinic acid mono-methyl ester. CO_2 is an antitranspirant that is reliable. A small rise in the concentration of CO_2 from the normal 0.03 per cent to 0.05 per cent causes partial stomata closure. A higher concentration cannot be used because this results in full stomata closure that adversely affects photosynthesis and breathing. PMA is a chemical used extensively for near stomatal application. It has the potential to close the stomata but it has the downside of being poisonous to fruits and vegetables. This chemical also prevents phosphorylation by blocking photosynthesis and thus destroys the foliage.

Reflectance Compound

White material that reflects solar radiation and when applied on the surface of the leaves increases the leaf albedo. Reflective compounds do not cause stomatal pores to block when applied exclusively on the lower surfaces to the upper surfaces of leaves with stomata. Chemical form reflectance coating decreases the temperature of the leaves. Experimentally proven we decreased a transpiration rate to 22-28 percent and also reduced leaf temperature from 3o to 4o after kaolinite coating (225 mg dm-2). Many chemical compounds reduce the temperature of the leaf by reflecting the solar radiation that causes the transpiration rate to slow and increase crop productivity in the use of water.

Disadvantages of Antitranspirants

An attractive prospect which merits further investigation is the possibility of using antitranspirants on grass to reduce both irrigation frequency and mowing. A promising area for study and studies is the use of antitranspirants to minimize transpirational water losses from shrubs and trees on watersheds, where improved water yields can be more important than any harm caused by growth reductions. Reductions in growth from the use of antitranspirants should not be disadvantageous once the oleanders have reached an effective height for screening headlight glare. Because of the low soil water potential and/or high evaporative demand, growth retarded growth is retarded by natural stomata1 closure when an untreated plant wilt. Through decreasing the rate at which water is lost, antitranspirants can help avoid wilting, or at least postpone it.

References

Davennport, D. C., P. E. Martin and M. R. Hagan (1969). Antitranspirants uses and effects on plant life. California Agriculture, 4:19.

Del Amor, F. M. and J. S. Rubio (2009). Effects of antitranspirant spray and potassium: calcium: magnesium ratio on photosynthesis, nutrient and water uptake, growth and yield of sweet pepper. Journal of Plant Nutrition, 32:97–111.

Evans, R. G. and E. J. Sadler (2008). Methods and technologies to improve efficiency of water use. Water Resources Research, 44: W00E04.

Faten, S. Abd El-Aal, M. Mona, A. Mouty and A. H. Ali (2008). Combined Effect of Irrigation Intervals and Foliar Application of Some Antitranspirants on Eggplant Growth, Fruits Yield and its Physical and Chemical Properties. Research Journal of Agriculture and Biological Sciences, 4(5) : 416- 423.

Gaballah, M. S., S. M. Shaaban and E. F. Abdallah (2014). The use of antitranspirants and organic compost in sunflower grown under water stress and sandy soil. International Journal of Academic Research, 6.